Lecture Notes in Mathematics

Edited by A. Dold and B. Eckmann

442

Calvin H. Wilcox

Scattering Theory for the d'Alembert Equation in Exterior Domains

Springer-Verlag
Berlin · Heidelberg · New York 1975

Dr. Calvin H. Wilcox
Department of Mathematics
The University of Utah
Salt Lake City, UT 84112/USA

Library of Congress Cataloging in Publication Data

Wilcox, Calvin Hayden.
Scattering theory for the d'Alembert wave equation in exterior domains.

(Lecture notes in mathematics ; 442)
Bibliography: p.
Includes index.
1. Sound-waves--Scattering. 2. Wave equation.
3. Lagrange equations. 4. Laplacian operator.
I. Title. II. Series: Lecture notes in mathematics (Berlin) ; 442.
QA3.L28 no.442 [QC243.3.S3] 510'.8s [534.2]
75-6605

AMS Subject Classifications (1970): 35B40, 35L05, 35P10, 35P25, 76Q05

ISBN 3-540-07144-X Springer-Verlag Berlin · Heidelberg · New York
ISBN 0-387-07144-X Springer-Verlag New York · Heidelberg · Berlin

Offsetdruck: Julius Beltz, Hemsbach/Bergstr.

CONTENTS

PREFACE

These lecture notes are the written version of a series of lectures given at Tulane University during the spring semester of 1974 and, in expanded form, at the University of Stuttgart during the summer semester of 1974. The goal of the lectures was to present a complete and self-contained exposition of the mathematical theory of scattering for a simple, but typical, wave propagation problem of classical physics. The problem selected for this purpose was that of the scattering of acoustic waves by a bounded rigid obstacle immersed in a homogeneous fluid. When formulated mathematically the problem becomes an initial-boundary value problem for the d'Alembert wave equation in an exterior domain. The lecture notes present a simple approach to this problem based on a selfadjoint extension of the Laplacian in Hilbert space. The principal results presented in the notes are the construction of eigenfunction expansions for the Laplacian and the calculation of the asymptotic form of solutions of the d'Alembert equation for large values of the time parameter.

The theory developed in the notes is an exposition and synthesis of results developed by several authors during a period of more than twelve years, unified and extended by a number of new results due to the author. A discussion of the related literature is given in Lecture

1. The new results which are presented here for the first time include the results on asymptotic wave functions for the d'Alembert equation (Lectures 2 and 8), the direct proof of the existence and structure of the wave operators based on the eigenfunction expansion (Lecture 7) and the proofs of the limiting absorption principle and eigenfunction expansion theorem for domains with non-smooth boundaries (Lectures 4 and 6). One interesting feature of the method developed here is that it makes no use of coercivity or elliptic estimates near the boundary for the Laplacian. In fact these concepts are not even mentioned in the lecture notes.

Before preparing these notes the author benefited from a number of discussions with Dr. William C. Lyford concerning eigenfunction expansions and scattering theory in domains with non-smooth boundaries. It is a pleasure to acknowledge his assistance.

The author would like to thank Professors E. D. Conway, J. A. Goldstein and S. I. Rosencrans of Tulane University and Professor Peter Werner of the University of Stuttgart for the opportunity to present these lectures at their universities. The lectures at Tulane were supported by a grant from the Ford Foundation. Moreover, the preparation of the written version of the lecture notes was supported in part by the U. S. Office of Naval Research. This support is hereby gratefully acknowledged.

Calvin H. Wilcox

LECTURE 1. INTRODUCTION

These lectures deal with the physical problem of the scattering of acoustic waves by a bounded rigid obstacle Γ immersed in an unlimited homogeneous fluid. It is assumed that a small-amplitude perturbation of the fluid exists at time $t = 0$ (due, for example, to forces acting during $t < 0$). The basic physical problem is to predict the evolution of the resulting acoustic wave during $t > 0$. This problem is solved below for arbitrary initial states with finite energy and a class of obstacles with irregular (non-smooth) surfaces. The class of allowable obstacles includes all of the simple, but non-smooth, surfaces that arise in applications, such as polyhedra, finite sections of cylinders, cones, spheres, disks, etc. One of the principal results of the analysis is that every wave with finite energy is asymptotically equal for $t \to \infty$ to a diverging spherical wave. Moreover, it is shown how the profile of this wave can be calculated from the initial state. These results are then used to calculate the asymptotic distribution of the energy for $t \to \infty$.

The following notation will be used in the mathematical formulation of acoustic wave propagation problem. $\mathbb{R}$ denotes the field of real numbers, $\mathbb{R}^n = \mathbb{R} \times \mathbb{R} \times \dots \times \mathbb{R}$ (n factors), $t \in \mathbb{R}$ and $x = (x_1, x_2, \dots, x_n) \in \mathbb{R}^n$. $\Omega \subset \mathbb{R}^n$ denotes an exterior domain; that is, Ω

is an open connected subset of $\mathbb{R}^n$ and $\Gamma = \mathbb{R}^n - \Omega$ is bounded.

The acoustic wave propagation problem, when formulated mathematically, leads to the following initial-boundary value problem for the d'Alembert wave equation [9]. A function

$$u : \mathbb{R} \times \Omega \to \mathbb{R} \tag{1.1}$$

is sought such that

$$D_0^2 u - (D_1^2 u + D_2^2 u + \dots + D_n^2 u) = 0 \quad \text{for} \quad t \in \mathbb{R},\ x \in \Omega, \tag{1.2}$$

$$D_\nu u \equiv \nu_1 D_1 u + \nu_2 D_2 u + \dots + \nu_n D_n u = 0 \quad \text{for } t \in \mathbb{R},\ x \in \partial\Omega, \tag{1.3}$$

$$u(0,x) = f(x) \quad \text{and} \quad D_0 u(0,x) = g(x) \quad \text{for} \quad x \in \Omega. \tag{1.4}$$

Here $D_0 = \partial/\partial t$, $D_j = \partial/\partial x_j$ for $j = 1,2,\dots,n$, $\partial\Omega$ represents the boundary of Ω and $\nu = (\nu_1, \nu_2, \dots, \nu_n)$ represents a unit vector normal to $\partial\Omega$ at x. The functions $f(x)$ and $g(x)$ are prescribed real-valued functions on Ω.

These equations and variables have the following physical interpretation for the acoustic wave propagation problem. n is the space-dimension ($n = 1,2$ or 3 in the applications). Ω represents the homogeneous fluid with sound speed $c = 1$ and $\Gamma = \mathbb{R}^n - \Omega$ represents the scattering obstacle. The function $u(t,x)$ is the acoustic potential. Thus $\vec{v} = \nabla u = (D_1 u, D_2 u, \dots, D_n u)$ represents the fluid velocity and $p = D_0 u$ represents the excess pressure of the acoustic disturbance. The d'Alembert wave equation

(1.2) is a consequence of the linearized equations of fluid dynamics; see, e.g. [9]. The Neumann boundary condition (1.3) describes the condition that Γ is rigid. The prescribed functions $f(x)$ and $g(x)$ represent the initial state of the acoustic field.

If $K \subset \Omega$ then the quantity

$$(1.5) \qquad E(u,K,t) = \int_K \sum_{k=0}^{n} \{D_k u(t,x)\}^2 \, dx,$$

where $dx = dx_1 dx_2 \ldots dx_n$, is interpreted as the acoustic energy in the set K at time t. In particular, solutions of (1.2), (1.3) satisfy the principle of conservation of energy:

$$(1.6) \qquad E(u,\Omega,t) = E(u,\Omega,0) = \text{const.}, \qquad t \in \mathbb{R},$$

where the constant may be finite or infinite. These lectures are concerned primarily with solutions with finite energy or "solutions wFE" as they will be called for brevity. The primary goal of these lectures is to study the asymptotic behavior for $t \to \infty$ of these solutions and the associated theory of scattering. In particular, the following topics are treated.

THE TRANSIENCY OF THE ENERGY IN BOUNDED SETS. Physical intuition suggests that if u is a solution wFE in an exterior domain Ω then

$$(1.7) \qquad \lim_{t\to\infty} E(u, K\cap\Omega, t) = 0 \quad \text{if } K \text{ is bounded;}$$

that is, the energy ultimately propagates out of every bounded set. A proof of this property is given in Lecture 5.

ASYMPTOTIC WAVE FUNCTIONS. Each initial state wFE satisfies

$$(1.8)\qquad E(u,\Omega,0) = \int_{\Omega}\left(\sum_{k=1}^{n} \{D_k f(x)\}^2 + g(x)^2\right)dx < \infty.$$

Such states are "quasi-localized" in the sense that to each $\epsilon > 0$ there corresponds a radius $R = R(\epsilon)$ such that

$$(1.9)\qquad E(u,\Omega,0) - \epsilon < E(u, B(R) \cap \Omega, 0) < E(u,\Omega,0)$$

where

$$(1.10)\qquad B(R) = \{x : |x| < R\}.$$

It follows that at any time $t_0 > 0$ the energy is contained in $B(R + t_0) \cap \Omega$, apart from a wave of total energy less than ϵ. Moreover, the energy propagates outward, by (1.7). This suggests that the velocity $\vec{v} = (D_1u, D_2u, \ldots, D_nu)$ and pressure $p = D_0u$ behave asymptotically like a diverging spherical wave:

$$(1.11)\qquad \begin{cases} D_k u(t,x) \sim u_k^{\infty}(t,x), \quad t \to \infty, \quad \text{where} \\ u_k^{\infty}(t,x) \equiv |x|^{\frac{1-n}{2}} F_k(|x| - t, x/|x|), \end{cases}$$

in the sense that, for $k = 0, 1, 2, \ldots, n$,

$$(1.12)\qquad \lim_{t\to\infty} \int_{\Omega} \{D_k u(t,x) - u_k^{\infty}(t,x)\}^2\, dx = 0.$$

Functions $u_k^{\infty}(t,x)$ of the form defined in (1.11) will be called "asymptotic wave functions". It is shown below that each solution wFE has unique asymptotic wave functions such that (1.12) holds. Moreover, the "wave profiles" $F_k(r,\eta)$, with $r \in \mathbb{R}$, $\eta \in S^{n-1} = \{x \in \mathbb{R}^n : |x| = 1\}$, are calculated from the initial state f,g.

ASYMPTOTIC ENERGY DISTRIBUTIONS. Consider a cone

$$C = \{x = r\eta : r > 0, \quad \eta \in C_0 \subset S^{n-1}\}. \tag{1.13}$$

Properties (1.7), (1.11), (1.12) imply that the fraction of the total energy $E(u,\Omega,0)$ which is contained in C at time t tends to a limit as $t \to \infty$. More precisely,

$$\lim_{t\to\infty} E(u, C\cap\Omega, t) = \int_{\mathbb{R}} \int_{C_0} \sum_{k=0}^{n} \{F_k(r,\eta)\}^2 \, d\eta \, dr \tag{1.14}$$

where $d\eta$ denotes the element of area on S^{n-1}. This behavior is verified below and the ultimate energy distribution (1.14) is calculated as a function of the initial state f,g.

SCATTERING THEORY. The time-dependent theory of scattering deals with the asymptotic equality of two systems for $t \to \infty$. Consider two exterior domains, Ω_1 and Ω_2, and corresponding solutions wFE, $u_1(t,x)$ and $u_2(t,x)$. The solutions will be said to be "asymptotically equal in energy" for $t \to \infty$ if

$$\lim_{t\to\infty} E(u_1 - u_2, \Omega_1 \cap \Omega_2, t) = 0. \tag{1.15}$$

Note that if $u_1(t,x)$ and $u_2(t,x)$ are asymptotically equal to the same solution wFE $u_0(t,x)$ in $\mathbb{R}^n$ (no obstacle) then (1.15) follows by the triangle inequality. Hence, to study asymptotic equality it is enough to compare solutions $u(t,x)$ in exterior domains Ω with solutions $u_0(t,x)$ in $\mathbb{R}^n$. Moreover, $u_0(t,x)$ will also have asymptotic wave functions, say

$$(1.16) \qquad \begin{cases} D_k u_0(t,x) \sim u_{k0}^{\infty}(t,x), \quad t \to \infty, \quad \text{where} \\ u_{k0}^{\infty}(t,x) \equiv |x|^{\frac{1-n}{2}} F_{k0}(|x|-t, x/|x|). \end{cases}$$

Hence, the relation

$$(1.17) \qquad \lim_{t\to\infty} E(u-u_0, \Omega, t) = 0$$

will follow from (1.11) and (1.16) if the initial state f_0, g_0 for u_0 can be adjusted so that the profiles F_{k0} and F_k coincide. It is shown below that this is always possible. The initial states f_0, g_0 and f,g are related by a "wave operator" in the sense of the time-dependent theory of scattering.

The remaining lectures are organized as follows.

LECTURE 2. SOLUTIONS OF THE D'ALEMBERT EQUATION IN $\mathbb{R}^n$. Here solutions in $\mathbb{R}^n$ are constructed by means of the Plancherel theory and used to derive the asymptotic behavior (1.11), (1.12) for the special case of solutions WFE in $\mathbb{R}^n$ (no boundary).

LECTURE 3. SOLUTIONS OF THE D'ALEMBERT EQUATION IN ARBITRARY DOMAINS. In this lecture $\Omega \subset \mathbb{R}^n$ represents an arbitrary domain (= open connected subset). A selfadjoint operator A on the Hilbert space $L_2(\Omega)$ is defined by the negative Laplacian, acting on a domain of functions which satisfy the Neumann condition in a suitably generalized sense. The spectral theorem for A is then used to discuss "solutions in $L_2(\Omega)$" and "solutions wFE" of the initial-boundary value problem.

LECTURE 4. STEADY-STATE SCATTERING THEORY IN EXTERIOR DOMAINS AND THE LIMITING ABSORPTION PRINCIPLE. Here the steady-state scattering problem is formulated for the d'Alembert equation in exterior domains and the uniqueness and existence of solutions is proved. The existence theorem is a corollary of a "limiting absorption theorem" which is the principal result of this lecture. It states that the resolvent $R_z = (A - z)^{-1}$ has limits, in a certain topology, when z tends to points of the spectrum of A. The proof of the theorem, and of most of the results in the subsequent lectures, are based on a version of the Rellich selection theorem [1,31] for the domain Ω. This is the only point in the theory where a restriction is imposed on the boundary of Ω. The Rellich theorem is established here for a class of domains called "domains with the finite tiling property". The class contains all the domains with piece-wise smooth boundaries having edges and corners that occur in applications, as well as many domains with highly singular boundaries.

LECTURE 5. TIME-DEPENDENT SCATTERING THEORY IN EXTERIOR DOMAINS. The lecture begins with F. Rellich's classical theorem that the operator A has no point spectrum. Then the limiting absorption theorem is shown to imply that A has no singular continuous spectrum. Next, the absolute continuity of the spectrum of A is shown to imply the transiency of energy in bounded sets, equation (1.7). Finally, the time-dependent scattering theory is formulated for $A^{\frac{1}{2}}$ and the corresponding operator $A_0^{\frac{1}{2}}$ in $L_2(\mathbb{R}^n)$. It is indicated that the existence, completeness and invariance of the wave operators can be derived from an abstract operator-theoretic theorem of M. S. Birman. This result is not needed in the development of scattering theory for the d'Alembert equation presented here, because a direct proof of the existence and completeness of the wave operators is given in Lecture 7. However, the proof based on the Birman theorem is presented in an Appendix because of its methodological interest.

LECTURE 6. STEADY-STATE SCATTERING THEORY AND EIGENFUNCTION EXPANSIONS FOR A. Here the limiting absorption theorem is used to construct generalized eigenfunctions for A, as solutions of a suitable steady-state scattering problem. Then an eigenfunction expansion theorem, in the sense of T. Ikebe's generalization of the Plancherel theory [14] is proved.

LECTURE 7. WAVE OPERATORS AND ASYMPTOTIC SOLUTIONS OF THE D'ALEMBERT EQUATION IN EXTERIOR DOMAINS. Here the eigenfunction expansion of Lecture 6 is used to construct solutions of the d'Alembert equation in Ω. Then a direct proof is given that these solutions are asymptotically equal to corresponding solutions in $\mathbb{R}^n$. This result leads to a direct proof of the existence and completeness of the wave operators, together with an explicit construction of them by means of the eigenfunctions of A and A_0.

LECTURE 8. ASYMPTOTIC WAVE FUNCTIONS AND ENERGY DISTRIBUTIONS IN EXTERIOR DOMAINS. Here the results of Lectures 2 and 7 are combined to construct the asymptotic wave functions $u_k^\infty(t,x)$ of (1.11) and prove their asymptotic equality to $D_k u(t,x)$. Finally, the asymptotic wave functions are used to calculate asymptotic energy distributions in various unbounded subsets of Ω.

The remainder of this lecture presents a brief discussion of the mathematical literature on which these lectures are based.

A theory of solutions wFE of the d'Alembert in arbitrary domains $\Omega \subset \mathbb{R}^n$ was given by the author in 1962 [43]. This work also dealt with more general hyperbolic equations having variable coefficients, and with solutions with locally finite energy. The initial-boundary value problem for hyperbolic equations in domains with more regular boundaries has a long history. A bibliography and discussion of this literature was given in [43]. The theory developed in [43] is based on the energy method. The approach based on the spectral theorem given in Lecture 3, which is available when the operator A is selfadjoint, is simpler than the energy method and leads to the same solutions wFE.

The limiting absorption theorem was first proved by D. M. Eidus in 1962 [7]. Since that time, his method has been applied by many authors to a variety of differential operators, boundary conditions and domains. An exposition of much of the Russian work was given by Eidus in [8]. The version given here for domains with irregular boundaries, subject to the "finite tiling condition" was suggested by work of W. C. Lyford [24,25].

The abstract theory of wave operators described in Lecture 5 and the Appendix is due to A. L. Belopolskii and M. S. Birman [2]. Their main theorem [2, Theorem 4.4] has been used by them [3,4] and other authors [24,25,34] to derive a number of abstract existence and completeness criteria for wave operators which are useful in different areas of application. The result used here, which is particularly well suited for applications to boundary value problems, is due to W. C. Lyford [25].

A different abstract theory of scattering has been developed by P. D. Lax and R. S. Phillips [19,20,21] and applied to the d'Alembert equation in exterior domains $\Omega \subset \mathbb{R}^n$ with n odd in [20] and with n even in [21]. In this work the boundary of Ω is assumed to be smooth ($\partial\Omega \in C^2$).

The results given in Lecture 6 generalize T. Ikebe's fundamental work on eigenfunction expansions for the Schrödinger operator in $\mathbb{R}^n$, $n \leq 3$ [14]. Ikebe's methods and results were first extended to the Laplacian in exterior domains by N. A. Shenk II [35] and to the Schrödinger equation in $\mathbb{R}^n$, $n \geq 4$ by D. W. Thoe [39]. A unified and extended version of these results is contained in a joint paper: N. A. Shenk II and D. W. Thoe [36]. Their results are based on the integral equation method of potential theory. More recently, A. J. Majda has given an alternative derivation [26], based on a method of Lax and Phillips [21], which does not use potential theory. In all the work cited above the boundary of Ω is assumed to be smooth ($\partial\Omega \in C^2$).

The theory of asymptotic wave functions for solutions in $\mathbb{R}^n$ of the d'Alembert equation, given in Lecture 2, is due to the author [45]. The generalization to solutions in exterior domains, given in Lecture 8, was indicated in [45]. Complete proofs are given here for the first time.

The versions of the limiting absorption theorem, eigenfunction expansion theorem and steady-state and time-dependent scattering theories given in the literature cited above all assume that $\partial\Omega \in C^2$. In essence, the generalization to non-smooth boundaries developed in these lectures is based on the discovery that "coercivity" or "elliptic estimates near the boundary" for the Laplacian in Ω are not needed in the theory. The only compactness result that is needed is the Rellich selection theorem. Rellich's theorem was proved by S. Agmon for domains with non-smooth boundaries which satisfy the "segment condition" [1, Theorem 3.8]. The version of Rellich's theorem presented here, based on the finite tiling condition, is a simple modification of Agmon's result.

LECTURE 2. SOLUTIONS OF THE D'ALEMBERT EQUATION IN $\mathbb{R}^n$

This lecture deals with the asymptotic behavior for $t \to \infty$ of solutions wFE of the d'Alembert equation in $\mathbb{R}^n$. The principal results are the calculation of the asymptotic wave functions and the proof of their convergence to the exact wave functions when $t \to \infty$. These are given in Theorems 2.8 and 2.10. The lecture begins with a brief discussion of the definition and construction of solutions of the d'Alembert equation in $\mathbb{R}^n$.

It will be convenient to introduce the Lebesgue space $L_2(\mathbb{R}^n)$ and several related spaces. By definition

$$\text{(2.1)} \quad \left\{ \begin{aligned} L_2(\mathbb{R}^n) = \{ [u] : u(x) \ &\text{is Lebesgue-measurable} \\ \text{on } \mathbb{R}^n,\ &\int_{\mathbb{R}^n} |u(x)|^2 dx < \infty\} \end{aligned} \right.$$

where $u(x)$ is a complex-valued function on $\mathbb{R}^n$ and

$$\text{(2.2)} \quad [u] = \{v(x) : v(x) = u(x) \quad \text{for almost all} \quad x \in \mathbb{R}^n\}$$

denotes the equivalence class of functions that differ from u on a Lebesgue null-set. It is well known that $L_2(\mathbb{R}^n)$ is a Hilbert space with scalar product

$$\text{(2.3)} \quad ([u],[v]) = \int_{\mathbb{R}^n} \overline{u(x)} v(x) dx$$

where $\overline{u(x)}$ denotes the complex conjugate of $u(x)$. It is natural to think of $L_2(\mathbb{R}^n)$ as a linear subspace of the Schwartz space $\mathcal{D}'(\mathbb{R}^n)$ of all distributions on $\mathbb{R}^n$. The embedding of $L_2(\mathbb{R}^n)$ into $\mathcal{D}'(\mathbb{R}^n)$ is given by

$$(2.4) \qquad [u](\phi) = \int_{\mathbb{R}^n} u(x)\phi(x)\,dx, \qquad \phi \in \mathcal{D}(\mathbb{R}^n).$$

From this point of view the Sobolev spaces for $\mathbb{R}^n$ may be defined by

$$(2.5) \quad L_2^m(\mathbb{R}^n) = L_2(\mathbb{R}^n) \cap \{[u] : D^\alpha[u] \in L_2(\mathbb{R}^n) \text{ for } |\alpha| \leq m\}.$$

Here m is a positive integer and the multi-index notation is used for derivatives. Thus $\alpha = (\alpha_1, \alpha_2, \ldots, \alpha_n)$ where each α_j is a non-negative integer, $D^\alpha = D_1^{\alpha_1} D_2^{\alpha_2} \cdots D_n^{\alpha_n}$, $D_j = \partial/\partial x_j$ and $|\alpha| = \alpha_1 + \alpha_2 + \ldots + \alpha_n$. $L_2^m(\mathbb{R}^n)$ is a Hilbert space with scalar product

$$(2.6) \qquad ([u],[v])_m = \int_{\mathbb{R}^n} \sum_{|\alpha|\leq m} \overline{D^\alpha u(x)} D^\alpha v(x)\,dx.$$

Similarly, the Laplacian $\Delta = D_1^2 + D_2^2 + \ldots + D_n^2$ can be used to define the Hilbert space

$$(2.7) \qquad L_2(\Delta, \mathbb{R}^n) = L_2(\mathbb{R}^n) \cap \{[u] : \Delta[u] \in L_2(\mathbb{R}^n)\}$$

with scalar product

$$(2.8) \qquad ([u],[v])_\Delta = ([u],[v]) + (\Delta[u], \Delta[v]).$$

In the remainder of these notes the equivalence class notation (2.2) is suppressed for notational simplicity. Thus elements of $L_2(\mathbb{R}^n)$ will be identified with functions on $\mathbb{R}^n$ and the scalar product formulas will be written

$$(2.9) \qquad (u,v) = \int_{\mathbb{R}^n} \overline{u(x)}\, v(x)\, dx .$$

A simple approach to integrating the d'Alembert equation in $\mathbb{R}^n$ is provided by the selfadjoint operator A_0 on $L_2(\mathbb{R}^n)$ corresponding to $-\Delta$. It is defined as the operator with domain of definition

$$(2.10) \qquad D(A_0) = L_2(\Delta, \mathbb{R}^n)$$

and action defined by

$$(2.11) \qquad A_0 u = -\Delta u \quad \text{for all} \quad u \in D(A_0) .$$

THEOREM 2.1. A_0 *is a selfadjoint non-negative operator on* $L_2(\mathbb{R}^n)$.

It follows that A_0 has a spectral family $\{\Pi_0(\lambda)\}$ and corresponding spectral representation

$$(2.12) \qquad A_0 = \int_0^\infty \lambda\, d\Pi_0(\lambda) .$$

Functions of A_0 can be defined by (see, e.g., [33])

$$(2.13) \qquad \Psi(A_0) = \int_0^\infty \Psi(\lambda)\, d\Pi_0(\lambda) .$$

They are, in general, unbounded operators with domain

$$(2.14) \quad D(\Psi(A_0)) = \{u : \int_0^\infty |\Psi(\lambda)|^2\, d\|\Pi_0(\lambda)u\|^2 < \infty\} .$$

In particular, the choice $\Psi(\lambda) = \lambda^{\frac{1}{2}} \geq 0$ gives

COROLLARY 2.2. A_0 *has a unique non-negative square root* $A_0^{\frac{1}{2}}$ *Moreover,* $D(A_0^{\frac{1}{2}}) = L_2^1(\mathbb{R}^n)$ *and*

$$(2.15)\qquad \|A_0^{\frac{1}{2}}u\|^2 = \sum_{j=1}^{n} \|D_j u\|^2 \quad \textit{for all} \quad u \in D(A_0^{\frac{1}{2}}).$$

Theorem 2.1 and Corollary 2.2 are simple consequences of the Plancherel theory. They are also special cases of Theorem 3.1 and Corollary 3.3 of Lecture 3. The proofs are therefore postponed to Lecture 3.

The construction (2.13) can be used to integrate the d'Alembert equation in $\mathbb{R}^n$. The d'Alembert equation is interpreted as the equation

$$(2.16)\qquad D_0^2 u_0 + A_0 u_0 = 0 \quad \text{for} \quad t \in \mathbb{R},$$

where u_0 is an $L_2(\mathbb{R}^n)$-valued function of $t \in \mathbb{R}$. If $u_0(t)$ has initial values in $L_2(\mathbb{R}^n)$,

$$(2.17)\qquad u_0(0) = f \quad \text{and} \quad D_0 u_0(0) = g \quad \text{in} \quad L_2(\mathbb{R}^n),$$

then

$$(2.18)\qquad u_0(t) = (\cos tA_0^{\frac{1}{2}})f + (A_0^{-\frac{1}{2}} \sin tA_0^{\frac{1}{2}})g$$

where the coefficient operators in (2.18) are bounded selfadjoint operators on $L_2(\mathbb{R}^n)$ defined by (2.13).

Another, more explicit, method for integrating the d'Alembert equation in $L_2(\mathbb{R}^n)$ is provided by the Plancherel theory of the Fourier transform [33]. The basic formulas may be written

$$(2.19)\qquad \begin{cases} \hat{f}(p) = (\Phi_n f)(p) \\[1ex] \quad = L_2(\mathbb{R}^n) - \lim\limits_{M \to \infty} \dfrac{1}{(2\pi)^{n/2}} \displaystyle\int_{|x| \le M} e^{-ix\cdot p} f(x)\,dx \end{cases}$$

and

$$
(2.20)\quad \begin{cases} f(x) = (\Phi_n^{*}\hat{f})(x) \\[1em] \quad = L_2(\mathbb{R}^n) - \lim\limits_{M\to\infty} \dfrac{1}{(2\pi)^{n/2}} \displaystyle\int_{|p|\le M} e^{ip\cdot x}\, \hat{f}(p)\,dp \end{cases}
$$

where $p = (p_1, p_2, \ldots, p_n) \in \mathbb{R}^n$ and $x \cdot p = x_1p_1 + x_2p_2 + \ldots + x_np_n$. The theory guarantees that the $L_2(\mathbb{R}^n)$-limit in (2.19) exists for all $f \in L_2(\mathbb{R}^n)$. The integral will not, in general, converge point-wise. Moreover, $\Phi_n : L_2(\mathbb{R}^n) \to L_2(\mathbb{R}^n)$ is a unitary operator with inverse $\Phi_n^{-1} = \Phi_n^{*}$ given by (2.20). In particular, Parseval's relation holds:

$$
(2.21)\qquad \|\hat{f}\| = \|f\| \quad \text{for all} \quad f \in L_2(\mathbb{R}^n).
$$

Φ_n provides a spectral representation of all differential operators with constant coefficients on $L_2(\mathbb{R}^n)$. This is a consequence of the relation

$$
(2.22)\qquad (\Phi_n D_j f)(p) = ip_j(\Phi_n f)(p), \qquad j = 1,2, \ldots, n,
$$

which is valid if f and D_jf are in $L_2(\mathbb{R}^n)$. The connection between the Fourier transform and A_0 is given by

$$
(2.23)\quad \begin{cases} \Psi(A_0)f(x) \\[1em] \quad = L_2(\mathbb{R}^n) - \lim\limits_{M\to\infty} \dfrac{1}{(2\pi)^{n/2}} \displaystyle\int_{|p|\le M} e^{ix\cdot p}\, \Psi(|p|^2)\hat{f}(p)\,dp. \end{cases}
$$

In particular, the spectral family $\Pi_0(\mu)$ corresponds

to $\Psi(\lambda) = H(\mu - \lambda)$ where $H(\tau)$ is Heaviside's function ($H(\tau) = 1$ for $\tau \geq 0$, $H(\tau) = 0$ for $\tau < 0$). Thus

$$(2.24) \quad \Pi_0(\mu) f(x) = \begin{cases} \dfrac{1}{(2\pi)^{n/2}} \displaystyle\int_{|p| \leq \sqrt{\mu}} e^{ix \cdot p} \hat{f}(p) dp, & \mu \geq 0, \\ 0 & , \mu < 0. \end{cases}$$

In what follows the cumbersome $L_2(\mathbb{R}^n)$-limit notation of (2.19), (2.20), (2.23) will be dropped. For example, (2.19) will henceforth be written in the symbolic form

$$(2.25) \qquad \hat{f}(p) = \frac{1}{(2\pi)^{n/2}} \int_{\mathbb{R}^n} e^{-ip \cdot x} f(x) dx.$$

However, the limit (2.19) must be restored when interpreting such formulas.

Application of the representation (2.23) to the solution (2.18) of the d'Alembert equation gives the Fourier integral form of the solution:

$$(2.26) \quad \begin{cases} u_0(t,x) = \dfrac{1}{(2\pi)^{n/2}} \displaystyle\int_{\mathbb{R}^n} e^{ix \cdot p} \Big\{ \hat{f}(p) \cos t|p| \\ \qquad\qquad\qquad\qquad + \hat{g}(p) \dfrac{\sin t|p|}{|p|} \Big\} dp. \end{cases}$$

Several remarks concerning the interpretation of the solution formulas (2.18) and (2.26) will be given next. It is clear that, in general, they do not define solutions of the d'Alembert equation in the classical sense. However, it is known that the initial value problem for the d'Alembert equation has a unique distribution solution for all initial values f, g in $\mathcal{D}'(\mathbb{R}^n)$; see [13] for references and further discussion. It is not difficult to show that (2.18) and (2.26) are meaningful and represent this unique distribution solution whenever f and g are in $L_2(\mathbb{R}^n)$. In this case $u_0(t)$ will be called the "solution in $L_2(\mathbb{R}^n)$" of the d'Alembert equation. It is easy to verify that it defines a continuous curve $t \to u_0(t) \in L_2(\mathbb{R}^n)$.

If $f \in D(A_0^{\frac{1}{2}}) = L_2^1(\mathbb{R}^n)$ and $g \in L_2(\mathbb{R}^n)$ then (2.18) implies that $u_0(t) \in L_2^1(\mathbb{R}^n)$ and $D_0 u_0(t) \in L_2(\mathbb{R}^n)$ and the curves $t \to D_j u_0(t) \in L_2(\mathbb{R}^n)$, $j = 0, 1, \ldots, n$ are continuous. In this case the energy integral is finite:

$$
(2.27)\quad \left\{
\begin{aligned}
E(u_0, \mathbb{R}^n, t) &= \int_{\mathbb{R}^n} \left\{ \sum_{j=1}^{n} |D_j f(x)|^2 + |g(x)|^2 \right\} dx \\
&= \int_{\mathbb{R}^n} \{ |p|^2 |\hat{f}(p)|^2 + |\hat{g}(p)|^2 \} dp,
\end{aligned}
\right.
$$

and u_0 is the "solution WFE in $\mathbb{R}^n$" of the d'Alembert equation. Finally, if $f \in D(A_0)$, $g \in D(A_0^{\frac{1}{2}}) = L_2^1(\mathbb{R}^n)$ then $u_0(t,x)$ will have first-order and second-order derivatives in $L_2(\mathbb{R}^n)$ and will satisfy the d'Alembert equation and initial conditions. In this case $u_0(t,x)$ is called the "strict solution WFE". These lectures deal primarily with solutions WFE and solutions in $L_2(\mathbb{R}^n)$.

The remainder of this lecture deals with the asymptotic behavior of $u_0(t,x)$ for $t \to \infty$. The analysis begins with a representation theorem.

THEOREM 2.3. *Let* f *and* g *be real-valued functions such that* $f \in L_2(\mathbb{R}^n)$ *and* $g \in D(A_0^{-\frac{1}{2}})$ *and define*

$$h = f + iA_0^{-\frac{1}{2}}g \in L_2(\mathbb{R}^n). \tag{2.28}$$

Then the solution in $L_2(\mathbb{R}^n)$ *defined by* (2.18) *satisfies*

$$u_0(t,x) = \mathrm{Re}\{v_0(t,x)\} \tag{2.29}$$

where $v_0(t,x)$ *is the complex-valued solution in* $L_2(\mathbb{R}^n)$ *of the d'Alembert equation defined by*

$$v_0(t,\cdot) = e^{-itA_0^{\frac{1}{2}}} h. \tag{2.30}$$

PROOF. Replace the cosine and sine in (2.18) by complex exponentials, using Euler's formula. This gives

$$u_0(t,x) = \tfrac{1}{2}v_0(t,x) + \tfrac{1}{2}w_0(t,x) \tag{2.31}$$

where $v_0(t,x)$ is given by (2.30) and

$$w_0(t,x) = e^{itA_0^{\frac{1}{2}}} (f - i\, A_0^{-\frac{1}{2}}g). \tag{2.32}$$

Now A_0 is a real operator on $L_2(\mathbb{R}^n)$; that is, it carries the complex conjugate of a function into the conjugate of the image: $A_0\bar{u} = \overline{A_0 u}$. It follows, since f and g are real-valued, that

$$\overline{w_0(t,\cdot)} = e^{-itA_0^{\frac{1}{2}}} (f + i\, A_0^{-\frac{1}{2}}g) = v_0(t,\cdot). \tag{2.33}$$

Combining (2.33) and (2.31) gives (2.29).

Theorem 2.3 implies that the asymptotic behavior of $u_0(t,x)$ is determined by that of $v_0(t,x)$. The Fourier representation of the last function, which follows from (2.23) and (2.30), is

$$(2.34)\qquad v_0(t,x) = \frac{1}{(2\pi)^{n/2}} \int_{\mathbb{R}^n} e^{i(x\cdot p - t|p|)}\, \hat{h}(p)\,dp$$

where

$$(2.35)\qquad \hat{h}(p) = \hat{f}(p) + i\,\frac{\hat{g}(p)}{|p|} \in L_2(\mathbb{R}^n).$$

The asymptotic behavior of $v_0(t,x)$ will be calculated first for the case where $\hat{h}$ is in the class

$$(2.36)\qquad \mathcal{D}_0(\mathbb{R}^n) = \mathcal{D}(\mathbb{R}^n) \cap \{\hat{h}(p) : \hat{h}(p) \equiv 0 \text{ for } |p| \le a,\quad a = a(\hat{h}) > 0\}$$

where $\mathcal{D}(\mathbb{R}^n)$ is the usual Schwartz space. The results will then be extended to all $h \in L_2(\mathbb{R}^n)$ by using the fact that $\{h : \hat{h} \in \mathcal{D}_0(\mathbb{R}^n)\}$ is dense in $L_2(\mathbb{R}^n)$.

The integral in (2.34) converges point-wise for functions $\hat{h} \in \mathcal{D}_0(\mathbb{R}^n)$. In fact if

$$(2.37)\qquad \operatorname{supp} \hat{h} \subset \{p : 0 < a \le |p| \le b\}$$

then

$$(2.38)\qquad v_0(t,x) = \frac{1}{(2\pi)^{n/2}} \int_{a\le|p|\le b} e^{i(x\cdot p - t|p|)}\, \hat{h}(p)\,dp.$$

Introduce spherical coordinates for p in (2.38):

$$(2.39)\qquad p = \rho\omega \quad \text{with } \rho \ge 0,\ \omega \in S^{n-1} \quad \text{and} \quad dp = \rho^{n-1} d\rho\, d\omega.$$

This gives the representation

$$(2.40)\quad \begin{cases} v_0(t,x) = \dfrac{1}{(2\pi)^{n/2}} \displaystyle\int_a^b \int_{S^{n-1}} e^{i\rho(x\cdot\omega - t)} \hat{h}(\rho\omega)\rho^{n-1} d\omega d\rho \\ \qquad = \dfrac{1}{(2\pi)^{n/2}} \displaystyle\int_a^b e^{-it\rho} V(x,\rho)\rho^{n-1} d\rho \end{cases}$$

where

$$(2.41)\qquad V(x,\rho) = \int_{S^{n-1}} e^{i\rho x\cdot\omega} \hat{h}(\rho\omega)d\omega.$$

It is easy to verify that $V(x,\rho)$ is a solution of the Helmholtz equation

$$(2.42)\qquad (\Delta + \rho^2)V(x,\rho) = 0 \quad \text{for all} \quad x \in \mathbb{R}^n.$$

The asymptotic behavior of $V(x,\rho)$ for $|x| \to \infty$ will be calculated and then used to find the behavior of $v_0(t,x)$ for $t \to \infty$. The form of the integral (2.41) suggests the use of the method of stationary phase. A precise version of the method has been developed for integrals of the form (2.41), but with S^{n-1} replaced by a smooth manifold S of dimension $n - 1$ in $\mathbb{R}^n$, by W. Littman [23] and M. Matsumura [28,29]. Their results, specialized to the integral (2.41), imply

THEOREM 2.4. *Let* $\hat{h} \in \mathcal{D}_0(\mathbb{R}^n)$ *satisfy* (2.37) *and define* $V(x,\rho)$ *and* $q_0(x,\rho)$ *by* (2.41) *and*

$$(2.43)\quad \begin{cases} V(x,\rho) = \left(\dfrac{2\pi}{i\rho|x|}\right)^{\frac{n-1}{2}} e^{i\rho|x|} \hat{h}(\rho\eta) \\ \qquad + \left(\dfrac{2\pi}{-i\rho|x|}\right)^{\frac{n-1}{2}} e^{-i\rho|x|}\hat{h}(-\rho\eta) + q_0(x,\rho) \end{cases}$$

respectively, where $x = |x|\eta$ *and* $\eta \in S^{n-1}$. *Then there exists a constant* $M_0 = M_0(h)$ *such that*

$$(2.44)\quad \begin{cases} |q_0(x,\rho)| \le \dfrac{M_0}{|x|^{\frac{n+1}{2}}} & \textit{for all } |x| > 0, \\ a \le \rho \le b \textit{ and } \eta \in S^{n-1}. \end{cases}$$

In (2.43) *the square roots are defined by the convention that if* $z = \pm i|z|$ *then* $z^{\frac{1}{2}} = \exp\{\pm\, i\pi/4\}|z|^{\frac{1}{2}}$ *with* $|z|^{\frac{1}{2}} \ge 0$.

The advantage of the Littman and Matsumura results over earlier versions of the stationary phase method is that the error estimate (2.44) holds uniformly for all $\eta \in S^{n-1}$ and $a \le \rho \le b$. Substitution of (2.43) into (2.40) gives

$$(2.45)\quad \begin{cases} v_0(t,x) = |x|^{\frac{1-n}{2}}\, G(|x| - t,\ x/|x|) \\ \qquad + |x|^{\frac{1-n}{2}}\, G'(|x| + t,\ x/|x|) + q_1(t,x) \end{cases}$$

where $G(r,\eta)$ and $G'(r,\eta)$ are the functions of $r \in \mathbb{R}$, $\eta \in S^{n-1}$ defined by

$$(2.46)\quad G(r,\eta) = \frac{1}{(2\pi)^{\frac{1}{2}}} \int_a^b e^{ir\rho}\, \hat{h}(\rho\eta)\, (-i\rho)^{\frac{n-1}{2}}\, d\rho,$$

$$(2.47)\quad G'(r,\eta) = \frac{1}{(2\pi)^{\frac{1}{2}}} \int_{-b}^{-a} e^{ir\rho}\, \hat{h}(\rho\eta)\, (-i\rho)^{\frac{n-1}{2}}\, d\rho$$

and

$$(2.48)\quad q_1(t,x) = \frac{1}{(2\pi)^{n/2}} \int_a^b e^{-i t\rho}\, q_0(x,\rho)\rho^{n-1}\, d\rho .$$

Combining (2.48) and the estimate (2.44) gives

$$(2.49)\quad \begin{cases} |q_1(t,x)| \le \dfrac{M_1}{|x|^{\frac{n+1}{2}}} & \text{for all } |x| > 0, \\ t \in \mathbb{R} \text{ and } \eta \in S^{n-1} \end{cases}$$

where $M_1 = (2\pi)^{-n/2}(b^n - a^n)M_0/n$.

Note that the first term on the right in (2.45) is a diverging spherical wave of the type introduced in Lecture 1. It will be shown that the two remaining terms tend to zero in $L_2(\mathbb{R}^n)$ when $t \to \infty$. Before doing this it will be convenient to extend the correspondence $h \to G$ defined by (2.46) to arbitrary $h \in L_2(\mathbb{R}^n)$. To this end note that for each $\hat{h} \in \mathcal{D}_0(\mathbb{R}^n)$ the function G defined by (2.46) is in $L_2(\mathbb{R}\times S^{n-1})$ and

$$(2.50)\quad \|G\|^2_{L_2(\mathbb{R}\times S^{n-1})} = \int_{a\le|p|\le b} |\hat{h}(p)|^2 dp = \|\hat{h}\|^2 = \|h\|^2 .$$

To verify this note that the Fourier transform of $G(r,\eta)$ with respect to r is $(-i\rho)^{\frac{n-1}{2}}\hat{h}(\rho\eta)\chi_{(a,b)}(\rho)$ where $\chi_{(a,b)}(\rho)$ is the characteristic function of $a \le \rho \le b$. Hence, an application of Parseval's formula gives

$$(2.51)\quad \begin{cases} \displaystyle\int_{\mathbb{R}} |G(r,\eta)|^2 dr = \int_a^b |\hat{h}(\rho\eta)|^2 \rho^{n-1}\, d\rho \\ \text{for all } \eta \in S^{n-1}. \end{cases}$$

Integration of (2.51) over $\eta \in S^{n-1}$ gives (2.50).

The relation (2.50) can be used to extend the mapping $h \to G \in L_2(\mathbb{R} \times S^{n-1})$ to all $h \in L_2(\mathbb{R}^n)$ by completion, because $\{h: \hat{h} \in \mathcal{D}_0(\mathbb{R}^n)\}$ is dense in $L_2(\mathbb{R}^n)$. However, a more explicit definition is needed in what follows. It is based on the observation that if $G \in L_2(\mathbb{R} \times S^{n-1})$ then Fubini's theorem implies that

$$(2.52) \quad \begin{cases} \|G\|^2_{L_2(\mathbb{R}\times S^{n-1})} = \displaystyle\int_{\mathbb{R}} \left(\int_{S^{n-1}} |G(r,\eta)|^2 d\eta \right) dr \\ \qquad = \displaystyle\int_{\mathbb{R}} \|G(r,\cdot)\|^2_{L_2(S^{n-1})} \, dr. \end{cases}$$

Moreover, (2.52) and the measurability statements of Fubini's theorem imply that $L_2(\mathbb{R} \times S^{n-1})$ is isomorphic, as a Hilbert space, with the Lebesgue space $L_2(\mathbb{R}, L_2(S^{n-1}))$ of functions $r \to G(r,\cdot) \in L_2(S^{n-1})$ which are square-integrable in the sense of Bochner [12]. In what follows, no special notation is used for this isomorphism. Instead a function $G(r,\eta)$ in one of these spaces will be identified with the corresponding function in the other space whenever it is convenient.

The Plancherel theory of the Fourier transform is valid in $L_2(\mathbb{R}, L_2(S^{n-1}))$. The defining formulas may be written

$$(2.53) \quad \hat{G}(\rho,\eta) = (\Phi_1 G)(\rho,\eta) = \lim_{M\to\infty} \frac{1}{(2\pi)^{\frac{1}{2}}} \int_{|r|\le M} e^{-i\rho r} G(r,\eta)\,dr,$$

(2.54) $$G(r,\eta) = (\Phi_1^* \hat{G})(r,\eta) = \lim_{M\to\infty} \frac{1}{(2\pi)^{\frac{1}{2}}} \int_{|\rho|\le M} e^{ir\rho}\hat{G}(\rho,\eta)d\rho,$$

where the convergence is in $L_2(\mathbb{R}, L_2(S^{n-1}))$ (or $L_2(\mathbb{R}\times S^{n-1})$) and

(2.55) $$\begin{cases} \|\hat{G}\|_{L_2(\mathbb{R}\times S^{n-1})} = \|G\|_{L_2(\mathbb{R}\times S^{n-1})} & \text{for all} \\ G \in L_2(\mathbb{R}\times S^{n-1}). \end{cases}$$

The definition of the wave profile $G \in L_2(\mathbb{R}\times S^{n-1})$ corresponding to an arbitrary $h \in L_2(\mathbb{R}^n)$ is based on the above remarks and the observation that, for any $h \in L_2(\mathbb{R}^n)$,

(2.56) $$\begin{cases} \|h\|^2 = \|\hat{h}\|^2 = \int_{\mathbb{R}^n} |\hat{h}(p)|^2 dp \\ \quad = \int_0^\infty \left(\int_{S^{n-1}} |(-i\rho)^{\frac{n-1}{2}} \hat{h}(\rho\eta)|^2 \, d\eta \right) d\rho. \end{cases}$$

It follows that, for each $h \in L_2(\mathbb{R}^n)$,

(2.57) $$\hat{G}(\rho,\eta) = \begin{cases} (-i\rho)^{\frac{n-1}{2}} \hat{h}(\rho\eta), & \rho \ge 0, \\ 0 \quad , & \rho < 0, \end{cases}$$

defines a function $\hat{G} \in L_2(\mathbb{R}\times S^{n-1})$. The notation $\hat{G}$ is used because $\hat{G}$ is the Fourier transform, in the sense of (2.53), of a unique function $G \in L_2(\mathbb{R}\times S^{n-1})$. This motivates the

DEFINITION. For each $h \in L_2(\mathbb{R}^n)$ the corresponding *wave profile* $G \in L_2(\mathbb{R} \times S^{n-1})$ is the function defined by (2.57) and (2.54). Moreover, the *asymptotic wave function* $v_0^\infty(t,x)$ corresponding to

$$(2.58) \qquad v_0(t,x) = e^{-itA_0^{\frac{1}{2}}} h(x), \quad h \in L_2(\mathbb{R}^n)$$

is defined by

$$(2.59) \qquad \begin{cases} v_0^\infty(t,x) = |x|^{\frac{1-n}{2}} G(|x| - t,\ x/|x|), \\ \qquad x \in \mathbb{R}^n - \{0\}, \quad t \in \mathbb{R} \end{cases}$$

where G is the wave profile for h.

Some basic properties of asymptotic wave functions of the form (2.59) are described by

THEOREM 2.5. *Let* $H \in L_2(\mathbb{R} \times S^{n-1})$ *and define*

$$(2.60) \quad w^\infty(t,x) = |x|^{\frac{1-n}{2}} H(|x|-t,\ x/|x|), \quad x \in \mathbb{R}^n - \{0\}, t \in \mathbb{R}.$$

Then

$$(2.61) \qquad w^\infty(t,\cdot) \in L_2(\mathbb{R}^n) \ \textit{for all} \quad t \in \mathbb{R}.$$

$$(2.62) \quad t \to w^\infty(t,\cdot) \in L_2(\mathbb{R}^n) \ \textit{is continuous for all} \quad t \in \mathbb{R}.$$

$$(2.63) \qquad \begin{cases} \|w^\infty(t,\cdot)\| \ \textit{is a monotone increasing function} \\ \textit{of} \ t \in \mathbb{R}. \end{cases}$$

$$(2.64) \qquad \lim_{t \to +\infty} \|w^\infty(t,\cdot)\| = \|H\|_{L_2(\mathbb{R} \times S^{n-1})}.$$

$$(2.65) \qquad \lim_{t\to-\infty} \|w^{\infty}(t,\cdot)\| = 0.$$

PROOF. The definition (2.60) and Fubini's theorem imply (2.61) and

$$(2.66) \quad \begin{cases} \|w^{\infty}(t,\cdot)\|^2 = \int_0^{\infty} dr \int_{S^{n-1}} |H(r-t,\eta)|^2 d\eta \\ \qquad\qquad\quad = \int_{-t}^{\infty} dr \int_{S^{n-1}} |H(r,\eta)|^2 d\eta. \end{cases}$$

This relation implies (2.63), (2.64) and (2.65). To verify (2.62) note that the Fourier transform of $H(r-t,\eta)$ in $L_2(\mathbb{R}, L_2(S^{n-1}))$ is $e^{-it\rho}\,\hat{H}(\rho,\eta)$. Thus, for all t and τ in $\mathbb{R}$,

$$(2.67) \quad \begin{cases} \|w^{\infty}(t,\cdot) - w^{\infty}(\tau,\cdot)\|^2 \\ \le \int_{-\infty}^{\infty} dr \int_{S^{n-1}} |H(r-t,\eta) - H(r-\tau,\eta)|^2 d\eta \\ = \int_{-\infty}^{\infty} d\rho \int_{S^{n-1}} |e^{it\rho} - e^{i\tau\rho}|^2 |\hat{H}(\rho,\eta)|^2 d\eta \end{cases}$$

by Parseval's formula. Now $|e^{it\rho} - e^{i\tau\rho}|^2 \le 4$ for all real t,τ and ρ and $\lim_{t\to\tau} |e^{it\rho} - e^{i\tau\rho}|^2 = 0$ for fixed ρ. Moreover, $\hat{H} \in L_2(\mathbb{R}\times S^{n-1})$. Hence, the last integral above tends to zero when $t \to \tau$ by Lebesgue's dominated convergence theorem. Thus, (2.67) implies (2.62).

The principal result of this lecture is

THEOREM 2.6. *For every* $h \in L_2(\mathbb{R}^n)$

$$\lim_{t\to+\infty} \|v_0(t,\cdot) - v_0^\infty(t,\cdot)\| = 0. \tag{2.68}$$

This result will be proved first for functions $\hat{h} \in \mathcal{D}_0(\mathbb{R}^n)$. The general case will then be verified by a density argument. The proof for $\hat{h} \in \mathcal{D}_0(\mathbb{R}^n)$ is based on a convergence lemma which will be used several times in the remaining lectures. The lemma will therefore be given in a more general form than is needed to prove Theorem 2.6, as follows.

LEMMA 2.7. *Let* $\Omega \subset \mathbb{R}^n$ *be an exterior domain and let* $u(t,x)$ *have the properties*

$$u(t,\cdot) \in L_2(\Omega) \quad \textit{for every} \quad t > t_0 , \tag{2.69}$$

$$\lim_{t\to\infty} \|u(t,\cdot)\|_{L_2(K\cap\Omega)} = 0 \quad \textit{for every compact } K \subset \mathbb{R}^n, \tag{2.70}$$

$$|u(t,x)| \le M/|x|^{\frac{n+1}{2}} \quad \textit{for every} \quad |x| > r_0 \tag{2.71}$$

where t_0, r_0 *and* M *are constants. Then*

$$\lim_{t\to\infty} \|u(t,\cdot)\|_{L_2(\Omega)} = 0. \tag{2.72}$$

PROOF. It can be assumed without loss of generality that $\partial\Omega \subset B(r_0) = \{x: |x| < r_0\}$. Then for all $r > r_0$ and $t > t_0$,

$$(2.73)\quad \begin{cases} \|u(t,\cdot)\|^2_{L_2(\Omega)} \\ = \|u(t,\cdot)\|^2_{L_2(B(r)\cap\Omega)} + \displaystyle\int_{|x|\ge r} |u(t,x)|^2 dx \\ \le \|u(t,\cdot)\|^2_{L_2(B(r)\cap\Omega)} + M^2 \displaystyle\int_{|x|\ge r} \frac{dx}{|x|^{n+1}} \\ \le \|u(t,\cdot)\|^2_{L_2(B(r)\cap\Omega)} + M^2 |S^{n-1}|/r \end{cases}$$

where $|S^{n-1}|$ is the hyper-area of S^{n-1}. This estimate follows from (2.69) and (2.71). Now make $t \to \infty$ with r fixed. (2.70) implies that

$$(2.74)\quad \overline{\lim_{t\to\infty}} \|u(t,\cdot)\|^2_{L_2(\Omega)} \le M^2 |S^{n-1}|/r \quad \text{for every } r > r_0.$$

This result implies (2.72) because the left-hand side of (2.74) is independent of r.

PROOF OF (2.68) FOR $\hat{h} \in \mathcal{D}_0(\mathbb{R}^n)$. Lemma 2.7, with $\Omega = \mathbb{R}^n$, will be applied to the difference

$$(2.75)\qquad q_2(t,x) = v_0(t,x) - v_0^\infty(t,x).$$

This can also be written

$$(2.76)\quad q_2(t,x) = |x|^{\frac{1-n}{2}} G'(|x|+t,\ x/|x|) + q_1(t,x)$$

by (2.45) and (2.59). Equation (2.75) and Theorem 2.5 imply that $q_2(t,\cdot) \in L_2(\mathbb{R}^n)$ for $t \ne 0$; i.e., (2.69) holds. Equation (2.70) will be verified for v_0 and v_0^∞ seperately. Note first that (2.34) with $\hat{h} \in \mathcal{D}_0(\mathbb{R}^n)$

implies (2.38) and hence

$$(2.77)\quad v_0(t,x) = \frac{1}{it} \int_{a\le|p|\le b} e^{-it|p|} \frac{\partial}{\partial|p|} \{e^{ip\cdot x}\, \hat{h}(p)\}dp.$$

Moreover, the integrand is uniformly bounded in t and hence

$$(2.78)\quad |v_0(t,x)| \le \frac{C}{|t|} \quad \text{for all} \quad x \in \mathbb{R}^n \quad \text{and} \quad t \neq 0$$

where C is a suitable constant. It follows that $v_0(t,\cdot)$ satisfies (2.70). As for $v_0^\infty(t,\cdot)$, note that (2.59) implies

$$(2.79)\quad \begin{cases} \|v_0^\infty(t,\cdot)\|^2_{L_2(B(R))} = \displaystyle\int_0^R \int_{S^{n-1}} |G(r-t,\eta)|^2 d\eta dr \\ \qquad\qquad = \displaystyle\int_{-t}^{R-t} \int_{S^{n-1}} |G(r',\eta)|^2 d\eta dr' \end{cases}$$

for all $R > 0$ and $t \neq 0$. Moreover, $G \in L_2(\mathbb{R} \times S^{n-1})$ and hence for R fixed the last integral tends to zero when $t \to \infty$. Hence $v_0^\infty(t,\cdot)$ satisfies (2.70). It follows from (2.75) and the triangle inequality that $q_2(t,\cdot)$ satisfies (2.70).

Equation (2.71) will be verified for $q_2(t,x)$ by means of (2.76). The estimate (2.71) was verified for $q_1(t,x)$ as (2.49). To prove it for $q_2(t,x)$ note that integration by parts in (2.47) implies an estimate

$$(2.80)\quad |G'(r,\eta)| \le \frac{C}{|r|} \quad \text{for all} \quad r \in \mathbb{R} - \{0\} \quad \text{and} \quad \eta \in S^{n-1}.$$

Hence

$$(2.81)\quad \begin{cases} \left| |x|^{\frac{1-n}{2}} G'(|x|+t, x/|x|) \right| \le \dfrac{C}{|x|^{\frac{n-1}{2}} (|x|+t)} \\[2ex] \le \dfrac{C}{|x|^{\frac{n+1}{2}}} \quad \text{for } |x|>0,\ t>0 \text{ and } \eta \in S^{n-1}. \end{cases}$$

Combining (2.49) and (2.81) gives (2.71) for $q_2(t,x)$. Thus $q_2(t,x)$ satisfies the conditions of Lemma 2.7 and hence (2.68) is proved for all $\hat{h} \in \mathcal{D}_0(\mathbb{R}^n)$.

PROOF OF THEOREM 2.6 - THE GENERAL CASE. Note that $v_0(t,\cdot) = U_0(t)h$ where

$$(2.82)\qquad U_0(t) = e^{-itA_0^{1/2}}$$

is a unitary operator on $L_2(\mathbb{R}^n)$. Similarly,

$$(2.83)\qquad v_0^\infty(t,\cdot) = U_0^\infty(t)h, \quad h \in L_2(\mathbb{R}^n)$$

defines a linear operator on $L_2(\mathbb{R}^n)$, by Theorem 2.5, and (2.63), (2.64) imply that

$$(2.84)\qquad \|U_0^\infty(t)\| \le 1 \quad \text{for all } t \in \mathbb{R}.$$

Now let h be any function in $L_2(\mathbb{R}^n)$ and let $\{h_m\}$ be a sequence such that $\hat{h}_m \in \mathcal{D}_0(\mathbb{R}^n)$ for $m = 1,2,3\ldots$ and $\lim_{m\to\infty} h_m = h$ in $L_2(\mathbb{R}^n)$. Then

$$(2.85)\quad \begin{cases} \|v_0(t,\cdot) - v_0^\infty(t,\cdot)\| = \|\{U_0(t) - U_0^\infty(t)\}h\| \\ \le \|\{U_0(t) - U_0^\infty(t)\}h_m\| + \|\{U_0(t) - U_0^\infty(t)\}(h - h_m)\| \\ \le \|U(t)h_m - U_0^\infty(t)h_m\| + 2\,\|h - h_m\| \end{cases}$$

for all $t > 0$ and $m = 1,2,3,\ldots$. Fix m and make $t \to \infty$ in (2.85). The first term on the right-hand side tends to zero, for any m, because $\hat{h}_m \in \mathcal{D}_0(\mathbb{R}^n)$. Thus

$$(2.86)\quad \overline{\lim_{t\to\infty}} \|v_0(t,\cdot) - v_0^\infty(t,\cdot)\| \le 2\,\|h - h_m\|, \quad m = 1,2,\ldots.$$

This implies (2.68) because the left-hand side is independent of m and $\lim_{m\to\infty} \|h - h_m\| = 0$.

Theorem 2.6 will now be used to calculate the asymptotic wave functions for solutions in $L_2(\mathbb{R}^n)$ and solutions WFE of the initial-value problem for the d'Alembert equation. For solutions in $L_2(\mathbb{R}^n)$ the result is stated as

THEOREM 2.8. *Let* f *and* g *be real-valued functions such that* $f \in L_2(\mathbb{R}^n)$ *and* $g \in D(A_0^{-\frac{1}{2}})$ *and let* $u_0(t,x)$ *be the corresponding solution in* $L_2(\mathbb{R}^n)$ *of the d'Alembert equation, given by* (2.26). *Define the asymptotic wave function*

$$(2.87)\quad u_0^\infty(t,x) = |x|^{\frac{1-n}{2}} F(|x|-t, x/|x|), \quad x \in \mathbb{R}^n - \{0\},\ t \in \mathbb{R},$$

by

$$(2.88)\quad F(r,\eta) = \mathrm{Re}\{G(r,\eta)\}$$

where $G(r,\eta)$ *is the complex-valued wave profile in* $L_2(\mathbb{R}\times S^{n-1})$ *defined by* (2.57) *with* $h = f + i\,A_0^{-\frac{1}{2}}g \in L_2(\mathbb{R}^n)$. *Then*

$$\lim_{t\to\infty} \|u_0(t,\cdot) - u_0^\infty(t,\cdot)\| = 0. \tag{2.89}$$

PROOF. The definitions (2.59) and (2.87), (2.88) imply

$$u_0^\infty(t,x) = \operatorname{Re}\{v_0^\infty(t,x)\}. \tag{2.90}$$

It follows from (2.29) and (2.90) that $u_0(t,x) - u_0^\infty(t,x) = \operatorname{Re}\{v_0(t,x) - v_0^\infty(t,x)\}$ and hence

$$\|u_0(t,\cdot) - u_0^\infty(t,\cdot)\| \le \|v_0(t,\cdot) - v_0^\infty(t,\cdot)\| \tag{2.91}$$

for all $t \in \mathbb{R}$. Thus Theorem 2.6 implies (2.89).

COROLLARY 2.9. *The asymptotic wave profile is characterized by the formulas*

$$(2.92)\quad \begin{cases} \hat{F}(\rho,\eta) = \frac{1}{2}(-i\rho)^{\frac{n-1}{2}} \begin{Bmatrix} \hat{h}(\rho\eta), & \rho > 0 \\ \overline{\hat{h}(-\rho\eta)}, & \rho < 0 \end{Bmatrix} \\ \qquad\quad = \frac{1}{2}(-i\rho)^{\frac{n-1}{2}} \left\{\hat{f}(\rho\eta) + i\,\dfrac{\hat{g}(\rho\eta)}{\rho}\right\}. \end{cases}$$

PROOF. The Fourier transform of a real-valued function $F(r,\eta)$ has the property

$$\hat{F}(\rho,\eta) = \overline{\hat{F}(-\rho,\eta)} \quad \text{for all } \rho \in \mathbb{R},\ \eta \in S^{n-1}. \tag{2.93}$$

The first equation in (2.92) follows from (2.93) and

the definitions (2.57) and (2.88). The second equation in (2.92) follows from the first, the definition of h and the properties $\hat{f}(-p) = \overline{\hat{f}(p)}$, $\hat{g}(-p) = \overline{\hat{g}(p)}$.

The asymptotic behavior for $t \to \infty$ of solutions wFE is considered next. Recall that if the initial state has finite energy, that is, if $f \in L_2^1(\mathbb{R}^n)$, $g \in L_2(\mathbb{R}^n)$, then the solution $u_0(t,x)$ has finite (constant) energy for all $t \in \mathbb{R}$. In particular, the first-order derivatives $D_k u_0(t,x)$, $k = 1,2,\ldots,n$, define continuous curves in $L_2(\mathbb{R}^n)$. It is not difficult to show that they are given by the following analogues of (2.26):

$$(2.94)\quad \begin{cases} D_0 u_0(t,x) = \dfrac{1}{(2\pi)^{n/2}} \displaystyle\int_{\mathbb{R}^n} e^{ix\cdot p}\{\hat{g}(p) \cos t|p| \\ \qquad\qquad - |p|\hat{f}(p) \sin t|p|\}dp \end{cases}$$

and

$$(2.95)\quad \begin{cases} D_k u_0(t,x) = \dfrac{1}{(2\pi)^{n/2}} \displaystyle\int_{\mathbb{R}^n} e^{ix\cdot p} \frac{ip_k}{|p|}\{|p|\hat{f}(p) \cos t|p| \\ \qquad\qquad + \hat{g}(p) \sin t|p|\}dp \end{cases}$$

for $k = 1,2, \ldots, n$. The spectral theorem for A_0 also implies the following representations:

$$(2.96)\quad D_0 u_0(t,\cdot) = (\cos tA_0^{\frac{1}{2}})g - (\sin tA_0^{\frac{1}{2}})A_0^{\frac{1}{2}}f$$

and

$$(2.97)\quad \begin{cases} D_k u_0(t,\cdot) = (\cos tA_0^{\frac{1}{2}})f_k + (\sin tA_0^{\frac{1}{2}})g_k, \\ k = 1,2, \ldots, n, \end{cases}$$

where f_k and g_k are the functions in $L_2(\mathbb{R}^n)$ defined by

$$(2.98)\qquad \begin{cases} \hat{f}_k(p) = ip_k\hat{f}(p), \quad \hat{g}_k(p) = i\dfrac{p_k}{|p|}\hat{g}(p), \\ \text{for } k = 1,2, \ldots, n. \end{cases}$$

The representations (2.96), (2.97) show that the first-order derivatives of a solution WFE are themselves solutions in $L_2(\mathbb{R}^n)$. Thus their asymptotic behavior can be derived from Theorem 2.8. The result is stated as

THEOREM 2.10. *Let* f *and* g *be real-valued functions such that* $f \in D(A_0^{\frac{1}{2}}) = L_2^1(\mathbb{R}^n)$ *and* $g \in L_2(\mathbb{R}^n)$ *and let* $u_0(t,x)$ *be the corresponding solution* WFE *of the d'Alembert equation. Define the asymptotic wave functions* $u_{k0}^{\infty}(t,x)$, $k = 0,1,2, \ldots, n$ *by*

$$(2.99)\qquad \begin{cases} u_{k0}^{\infty}(t,x) = |x|^{\frac{1-n}{2}} F_k(|x| - t, x/|x|), \\ x \in \mathbb{R}^n - \{0\}, \quad t \in \mathbb{R}, \end{cases}$$

$$(2.100)\qquad F_k(r,\eta) = \mathrm{Re}\{G_k(r,\eta)\}$$

$$(2.101)\qquad \hat{G}_k(\rho,\eta) = \begin{cases} (-i\rho)^{\frac{n-1}{2}}\hat{h}_k(\rho\eta), & \rho > 0 \\ 0 \quad , & \rho < 0 \end{cases}$$

and

$$(2.102)\quad \begin{cases} \hat{h}_0(p) = -i|p|\hat{f}(p) + \hat{g}(p) \\ \hat{h}_k(p) = \hat{f}_k(p) + i\hat{g}_k(p), \quad k=1,2,\dots,n. \end{cases}$$

Then

$$(2.103)\quad \lim_{t\to\infty} \|D_k u_0(t,\cdot) - u_{k0}^{\infty}(t,\cdot)\| = 0, \quad k = 0,1,2,\dots,n.$$

PROOF. Comparison of (2.18) and (2.96) shows that Theorem 2.8 is applicable to $D_0u_0(t,x)$ with f and $A_0^{-\frac{1}{2}}g$ replaced by g and $-A_0^{\frac{1}{2}}f$, respectively. This gives (2.103) with $k = 0$. Similarly, Theorem 2.8 is applicable to $D_ku_0(t,x)$ with $k = 1,2,\dots,n$ and with f and $A_0^{-\frac{1}{2}}g$ replaced by f_k and g_k, respectively, which gives (2.103) in the remaining cases.

LECTURE 3. SOLUTIONS OF THE D'ALEMBERT EQUATION IN ARBITRARY DOMAINS

In Lecture 2, the spectral theorem for the operator $A_0 = -\Delta$ on $L_2(\mathbb{R}^n)$ was used to give a simple construction of solutions wFE of the d'Alembert equation in $\mathbb{R}^n$. In this lecture the method is generalized to include the initial-boundary value problem (1.2), (1.3), (1.4) in arbitrary domains $\Omega \subset \mathbb{R}^n$. The resulting theory provides a foundation for the detailed study of the structure of solutions wFE in Lectures 4 - 8.

The formulation of the initial-boundary value problem given below is based on the following function spaces.

$$\text{(3.1)} \quad \begin{cases} L_2(\Omega) = \{u: u(x) \text{ is Lebesgue-measurable on } \Omega, \\ \qquad\qquad \int_\Omega |u(x)|^2 dx < \infty\} \end{cases}$$

$$\text{(3.2)} \quad L_2^m(\Omega) = L_2(\Omega) \cap \{u: D^\alpha u \in L_2(\Omega) \quad \text{for} \quad |\alpha| \le m\}$$

$$\text{(3.3)} \qquad L_2(\Delta,\Omega) = L_2(\Omega) \cap \{u: \Delta u \in L_2(\Omega)\}$$

$$\text{(3.4)} \qquad L_2^1(\Delta,\Omega) = L_2^1(\Omega) \cap L_2(\Delta,\Omega).$$

These spaces are Hilbert spaces with respect to the inner products

$$(3.5)\qquad (u,v) = \int_\Omega \overline{u(x)}\, v(x)\, dx$$

$$(3.6)\qquad (u,v)_m = \sum_{|\alpha|\le m} (D^\alpha u, D^\alpha v)$$

$$(3.7)\qquad (u,v)_\Delta = (u,v) + (\Delta u, \Delta v)$$

$$(3.8)\qquad (u,v)_{1,\Delta} = (u,v)_1 + (\Delta u, \Delta v)\,.$$

Recall that the elements of these spaces are actually equivalence classes of functions on Ω, although this is not indicated in the notation.

The formulation of the Neumann boundary condition for arbitrary domains Ω is motivated by Green's formula,

$$(3.9)\qquad \int_\Omega \{(\Delta u)v + \nabla u\cdot\nabla v\}dx = \int_{\partial\Omega} (\vec{v}\cdot\nabla u)v\,dS,$$

where $\nabla u = (D_1 u, D_2 u, \ldots, D_n u)$. (3.9) is valid whenever the boundary $\partial\Omega$ and the functions u and v are sufficiently regular. Moreover, if u satisfies the Neumann condition on $\partial\Omega$ then the right-hand side of (3.9) is zero for all v for which (3.9) holds. This suggests the

DEFINITION. A function $u \in L_2^1(\Delta,\Omega)$ is said to satisfy the *generalized Neumann condition* for Ω if and only if

$$(3.10)\qquad \int_\Omega \{(\Delta u)v + \nabla u\cdot\nabla v\}dx = 0 \quad \text{for all} \quad v \in L_2^1(\Omega)\,.$$

Note that (3.10) is meaningful for arbitrary domains Ω and defines a closed subspace

$$(3.11) \quad L_2^N(\Delta,\Omega) = L_2^1(\Delta,\Omega) \cap \{u: u \text{ satisfies } (3.10)\}$$

in the Hilbert space $L_2^1(\Delta,\Omega)$. Moreover, (3.10) is equivalent to the classical Neumann condition whenever $\partial\Omega$ is sufficiently smooth. For in this case functions $v \in L_2^1(\Omega)$ will have a trace on $\partial\Omega$ and the gradient ∇u of functions $u \in L_2^1(\Delta,\Omega)$ will have a trace on $\partial\Omega$, by the Sobolev imbedding theorems [22]. Moreover, (3.9) will hold. Comparison with (3.10) then gives

$$(3.12) \qquad \int_{\partial\Omega} (\vec{\nu} \cdot \nabla u) v dS = 0 \quad \text{for all} \quad v \in L_2^1(\Omega).$$

which implies that $\vec{\nu} \cdot \nabla u = 0$ on $\partial\Omega$. It is shown below that the initial-boundary value problem for the d'Alembert equation in an arbitrary domain Ω has a unique solution which satisfies (3.10). It follows from the remark above that this solution coincides with the classical solution whenever the latter exists.

The construction of the solution of the initial-boundary value problem is based on the linear operator $A: L_2(\Omega) \to L_2(\Omega)$ defined by

$$(3.13) \qquad D(A) = L_2^N(\Delta,\Omega),$$

$$(3.14) \qquad Au = -\Delta u \quad \text{for all} \quad u \in D(A).$$

The utility of this operator is based on

THEOREM 3.1. A *is a selfadjoint operator on the Hilbert space* $L_2(\Omega)$. *Moreover* $A \geq 0$.

The proof of this theorem is based on

LEMMA 3.2. *Let* $\mathcal{H}$ *be an abstract Hilbert space and let* $H: \mathcal{H} \to \mathcal{H}$ *be a linear operator whose domain* $D(H)$ *is dense in* $\mathcal{H}$. *Assume that*

$$H \subset H^* \tag{3.15}$$

$$H \geq 0 \tag{3.16}$$

$$R(1 + H) = \mathcal{H}. \tag{3.17}$$

Then H *is selfadjoint.*

PROOF OF LEMMA 3.2. The hypothesis that $D(H)$ is dense in $\mathcal{H}$ implies that H^*, the adjoint of H, is a well-defined linear operator. Condition (3.15) (the symmetry of H) means that H^* is an extension of H. It implies that (Hu,u) is real-valued for all $u \in D(H)$. Condition (3.16) means that $(Hu,u) \geq 0$ for all $u \in D(H)$. Condition (3.17) means that the range of $1 + H$ is $\mathcal{H}$, where 1 denotes the identity operator on $\mathcal{H}$.

The proof of Lemma 3.2 is based on standard results in the theory of linear operators in Hilbert space which can be found, for example, in [33]. Condition (3.16) implies that the deficiency indices of H are equal [18, p. 268] and then (3.17) implies that they are both zero. It follows that H is selfadjoint.

PROOF OF THEOREM 3.1. The conditions of Lemma 3.2 will be verified for $\mathcal{H} = L_2(\Omega)$ and $H = A$. It is clear that $\mathcal{D}(\Omega) \subset D(A)$ and hence $D(A)$ is dense in

$L_2(\Omega)$. To verify (3.15) let u_1 and u_2 be any two elements of $D(A)$. Then (3.10) holds with u_2 for u and $\bar{u}_1$ for v and also with $\bar{u}_1$ for u and u_2 for v. Thus

$$(3.18)\quad \begin{cases} (u_1, Au_2) = -\int_\Omega \bar{u}_1 \Delta u_2 dx = \int_\Omega \nabla \bar{u}_1 \cdot \nabla u_2 dx \\ \qquad = -\int_\Omega \overline{\Delta u_1} u_2 dx = (Au_1, u_2) \end{cases}$$

for all u_1 and u_2 in $D(A)$. It follows that $D(A) \subset D(A^*)$ and $A^*u = Au$ for all $u \in D(A)$; that is, $A \subset A^*$.

To verify (3.16) consider (3.18) with $u_1 = u_2 = u \in D(A)$. This gives

$$(3.19)\quad (u, Au) = \int_\Omega |\nabla u(x)|^2 dx \geq 0 \quad \text{for all} \quad u \in D(A)$$

which is equivalent to $A \geq 0$.

Only (3.17) remains to be verified; that is, $R(1 + A) = L_2(\Omega)$. Note that this is equivalent to the statement that for each $f \in L_2(\Omega)$ there exists a $u \in D(A) = L_2^N(\Delta, \Omega)$ such that $u + Au = f$, and hence

$$(3.20)\quad (u,v) + (Au,v) = (f,v) \quad \text{for all} \quad v \in L_2^1(\Omega).$$

Now (3.20) implies

$$(3.21)\quad \begin{cases} (u,v)_1 \equiv \int_\Omega \{\bar{u} v + \nabla \bar{u} \cdot \nabla v\} dx \\ \qquad = \int_\Omega \bar{f} v dx \quad \text{for all} \quad v \in L_2^1(\Omega) \end{cases}$$

because $u \in L_2^N(\Delta, \Omega)$. The proof of (3.17) will be completed by showing that (3.21) has a solution $u \in L_2^1(\Omega)$

for each $f \in L_2(\Omega)$ and then showing that (3.21) implies that $u \in L_2^N(\Delta,\Omega)$ and $u + Au = f$.

To prove that (3.21) always has a solution, let $f \in L_2(\Omega)$ and note that

$$(3.22) \quad |(f,v)| \le \|f\| \, \|v\| \le \|f\| \, \|v\|_1 \quad \text{for all } v \in L_2^1(\Omega).$$

Hence, the Riesz representation theorem in the Hilbert space $L_2^1(\Omega)$ implies that there exists a $u \in L_2^1(\Omega)$ such that (3.21) holds. To show that $\Delta u \in L_2(\Omega)$ and $\Delta u = u - f$ note that the definitions of Δu and ∇u in $\mathcal{D}'(\Omega)$ imply that

$$(3.23) \quad \Delta\bar{u}(v) = \int_\Omega \bar{u}\Delta v dx = -\int_\Omega \nabla\bar{u} \cdot \nabla v dx \quad \text{for all } v \in \mathcal{D}(\Omega).$$

Moreover, taking $v \in \mathcal{D}(\Omega) \subset L_2^1(\Omega)$ in (3.21) gives

$$(3.24) \quad \int_\Omega \nabla\bar{u} \cdot \nabla v dx = \int_\Omega (\bar{f} - \bar{u}) v dx \quad \text{for all } v \in \mathcal{D}(\Omega).$$

Combining (3.23) and (3.24) gives $\Delta u = u - f \in L_2(\Omega)$. This also implies that $u \in L_2^1(\Delta,\Omega)$.

Finally, to verify the boundary condition (3.10) combine (3.21) and the equation $f = u - \Delta u$. This gives

$$(3.25) \quad \int_\Omega \{\bar{u}v + \nabla\bar{u}\cdot\nabla v\} dx = \int_\Omega (\bar{u} - \Delta\bar{u}) v dx \quad \text{for all } v \in L_2^1(\Omega)$$

which is equivalent to (3.10). It follows that $u \in L_2^N(\Delta,\Omega)$ and $u + Au = f$, which completes the proof that A is selfadjoint. The property $A \ge 0$ was proved by (3.19) above.

The operator A has a non-negative square root $A^{\frac{1}{2}}$ whose domain is characterized by

COROLLARY 3.3. $D(A^{\frac{1}{2}}) = L_2^1(\Omega)$ *and*

$$(3.26)\qquad \|A^{\frac{1}{2}}u\|^2 = \sum_{j=1}^{n} \|D_j u\|^2 \quad \textit{for all} \quad u \in D(A^{\frac{1}{2}}).$$

PROOF. This result follows from the theory of sesquilinear forms in Hilbert space and the associated operators [18, Ch. VI]. In particular, A is the operator associated with the form $A(u,v) = (\nabla u, \nabla v)$ with domain $L_2^1(\Omega)$, and Corollary 3.3 is a special case of Kato's second representation theorem [18, p. 331].

The operator A will now be used to construct various classes of solutions of the initial-boundary value problem. First of all, the problem can be reformulated as an initial value problem in $L_2(\Omega)$. A function $u\colon \mathbb{R} \to L_2(\Omega)$ is sought such that

$$(3.27)\qquad D_0^2 u + Au = 0 \quad \text{for all} \quad t \in \mathbb{R}$$

$$(3.28)\qquad u(0) = f \quad \text{and} \quad D_0 u(0) = g \quad \text{in} \quad L_2(\Omega).$$

The spectral theorem for A:

$$(3.29)\qquad A = \int_0^\infty \lambda \, d\Pi(\lambda),$$

and the associated operator calculus make it possible to construct the generalized solution

$$(3.30)\qquad u(t) = (\cos tA^{\frac{1}{2}})f + (A^{-\frac{1}{2}} \sin tA^{\frac{1}{2}})g.$$

The coefficient operators in (3.30) are bounded and hence $u(t)$ is defined for all f and g in $L_2(\Omega)$ and defines a curve in $C(\mathbb{R}, L_2(\Omega))$, the class of continuous $L_2(\Omega)$-valued functions on $\mathbb{R}$. The differentiability properties of $u(t)$ depend on those of f and g. Three cases will be mentioned.

SOLUTIONS IN $L_2(\Omega)$. $f \in L_2(\Omega)$, $g \in L_2(\Omega)$. In this case $u(t)$ is continuous in $L_2(\Omega)$ and $u(0) = f$. However, $u(t)$ will not in general be differentiable, and hence (3.27) and the second initial condition need not hold. In this case $u(t)$ coincides with the "generalized solution in $L_2(\Omega)$" which was defined and studied by M. Vishik and O. A. Ladyzhenskaya [41].

SOLUTIONS WITH FINITE ENERGY. $f \in D(A^{\frac{1}{2}}) = L_2^1(\Omega)$, $g \in L_2(\Omega)$. In this case u is in the class

$$(3.31) \qquad C^1(\mathbb{R}, L_2(\Omega)) \cap C(\mathbb{R}, L_2^1(\Omega)).$$

This follows easily from the spectral theorem and Corollary 3.3. Hence, u satisfies (3.28), but (3.27) need not hold. In this case $u(t)$ coincides with the "solution wFE" which, for arbitrary domains Ω, was defined and studied by the author in [43]. The existence and uniqueness of solutions wFE was proved in [43] for arbitrary domains Ω and initial states $f \in L_2^1(\Omega)$, $g \in L_2(\Omega)$.

STRICT SOLUTIONS WITH FINITE ENERGY. $f \in D(A) = L_2^N(\Delta,\Omega)$, $g \in D(A^{\frac{1}{2}}) = L_2^1(\Omega)$. In this case u is in the class

$$(3.32) \quad C^2(\mathbb{R}, L_2(\Omega)) \cap C^1(\mathbb{R}, L_2^1(\Omega)) \cap C(\mathbb{R}, L_2^N(\Delta,\Omega))$$

and satisfies both (3.27) and (3.28). The existence of such "strict solutions wFE" was proved in [43] for arbitrary domains Ω and initial states $f \in L_2^N(\Delta,\Omega)$, $g \in L_2^1(\Omega)$.

Corollary 3.3 implies that the energy integral of the solution wFE can be written in the two equivalent forms

$$(3.33)\quad \begin{cases} E(u,\Omega,t) = \sum_{k=0}^{n} \|D_k u(t)\|^2 \\ \qquad = \|A^{\frac{1}{2}} u(t)\|^2 + \|D_0 u(t)\|^2 . \end{cases}$$

The constancy of the energy, and hence the equation,

$$(3.34)\quad E(u,\Omega,t) = E(u,\Omega,0) = \sum_{k=1}^{n} \|D_k f\|^2 + \|g\|^2$$

follows easily from the spectral theorem and the last equation of (3.33).

The analysis of the asymptotic behavior of $u(t)$ for $t \to \infty$, given in Lectures 5, 7 and 8, is based on the following generalization of Theorem 2.3.

THEOREM 3.4. *Let* f *and* g *be real-valued functions such that* $f \in L_2(\Omega)$ *and* $g \in D(A^{-\frac{1}{2}})$ *and define*

$$(3.35)\qquad h = f + i\, A^{-\frac{1}{2}} g \in L_2(\Omega).$$

Then the solution in $L_2(\Omega)$ *defined by* (3.30) *satisfies*

$$(3.36)\qquad u(t,x) = \mathrm{Re}\{v(t,x)\}$$

where $v(t,x)$ *is the complex-valued solution in* $L_2(\Omega)$ *of the d'Alembert equation defined by*

$$(3.37)\qquad v(t,\cdot) = e^{-itA^{\frac{1}{2}}} h.$$

The proof is exactly the same as that given for Theorem 2.3.

LECTURE 4. STEADY-STATE SCATTERING THEORY IN EXTERIOR DOMAINS AND THE LIMITING ABSORPTION PRINCIPLE

The acoustic wave $u(t,x)$ in Ω produced by a prescribed distribution of forces with potential function $f(t,x)$ is a solution of the inhomogeneous d'Alembert equation [9]

$$(4.1)\quad D_0^2 u - (D_1^2 u + D_2^2 u + \ldots + D_n^2 u) = f(t,x) \quad \text{for } t \in \mathbb{R},\ x \in \Omega.$$

This lecture is concerned with solving (4.1) in the case where the forces have a sinusoidal time dependence

$$(4.2)\qquad f(t,x) = e^{-i\omega t} f(x), \quad \omega \geq 0,$$

and solutions having the same time dependence are sought:

$$(4.3)\qquad u(t,x) = e^{-i\omega t} v(x).$$

This problem is called the steady-state scattering problem for the d'Alembert equation. The function $v(x)$ will be the solution of a boundary value problem which, in its classical formulation, is

$$(4.4)\qquad D_1^2 v + D_2^2 v + \ldots + D_n^2 v + \omega^2 v = -f(x) \quad \text{for } x \in \Omega,$$

$$(4.5)\qquad \begin{cases} D_\nu v(x) \equiv \nu_1 D_1 v(x) + \nu_2 D_2 v(x) + \ldots + \nu_n D_n v(x) = 0 \\ \text{for } x \in \partial\Omega. \end{cases}$$

Of course, the initial conditions (1.4) of Lecture 1 must be dropped because it is not known in advance which initial states will produce the desired time dependence (4.3).

Conditions (4.4), (4.5) are formally equivalent to the equation $(A - \omega^2)v = f$ where A is the selfadjoint operator defined in Lecture 3. Now $A \geq 0$ which implies that $\sigma(A)$, the spectrum of A, is a subset of $\overline{\mathbb{R}}_+ = \{\lambda : \lambda \geq 0\}$. In fact, it will be shown in Lecture 5 below that $\sigma(A) = \overline{\mathbb{R}}_+$. It follows that (4.4), (4.5) cannot have solutions v in $L_2(\Omega)$ and the solutions must be sought in a larger function space. Solutions in this space will not be unique unless additional conditions are imposed. The well-known Sommerfeld radiation condition will be used here. Physically, the condition implies that the wave (4.3) behaves like a diverging spherical wave when $|x| \to \infty$.

The existence of steady-state solutions will be proved by the method of limiting absorption. This is based on the observation that if $\zeta \in \mathbb{C}$ and $\operatorname{Im} \zeta \neq 0$ then

$$(4.6) \qquad (A - \zeta)v = f$$

has a solution $v(x,\zeta) \in L_2(\Omega)$ for each $f \in L_2(\Omega)$ because $\zeta \notin \sigma(A)$. The limiting absorption method is to seek the steady-state solution as the limit

$$(4.7) \qquad v(x,\omega) = \lim_{\zeta \to \omega^2} v(x,\zeta)$$

where the limit is sought not in $L_2(\Omega)$, but in a suitable larger function space. Physically, the function

$v(x,\zeta)$ with $\text{Im}\,\zeta \neq 0$ describes a steady-state wave in a lossy, or energy-absorbing, fluid with absorption coefficient proportional to $\text{Im}\,\zeta$ [42].

The validity of the limiting absorption method is proved below by means of a local compactness theorem which is essentially the Rellich selection theorem [1,31]. It is necessary to restrict the class of boundaries $\partial\Omega$ to guarantee the validity of this result. It is noteworthy that in the remainder of the lectures the class of boundaries $\partial\Omega$ (or obstacles Γ) is restricted only through the application of this local compactness theorem.

The remainder of this lecture is organized as follows. First several function spaces are introduced and used to formulate the steady-state scattering problem in arbitrary exterior domains Ω. Then the Sommerfeld radiation condition is introduced and the uniqueness theorem is proved for arbitrary exterior domains. The local compactness theorem is given next. Then the limiting absorption theorem is formulated and proved by means of the uniqueness and local compactness theorems. Finally, the existence of solutions to the steady-state scattering problem is obtained as a corollary of the limiting absorption theorem.

The formulation of the steady-state scattering problem makes use of several spaces of functions which are locally in the Hilbert spaces of Lecture 3. They may be defined as follows:

$$(4.8) \quad \begin{cases} L_2^{loc}(\bar{\Omega}) = \mathcal{D}'(\Omega) \cap \{u\colon u \in L_2(K \cap \Omega) \quad \text{for all} \\ \qquad\qquad\qquad\qquad\qquad\qquad \text{compact } K \subset \mathbb{R}^n\} \end{cases}$$

$$(4.9) \quad L_2^{m,loc}(\bar{\Omega}) = L_2^{loc}(\bar{\Omega}) \cap \{u\colon D^\alpha u \in L_2^{loc}(\bar{\Omega}) \quad \text{for} \quad |\alpha| \le m\}$$

$$(4.10) \qquad L_2^{loc}(\Delta,\bar{\Omega}) = L_2^{loc}(\bar{\Omega}) \cap \{u\colon \Delta u \in L_2^{loc}(\bar{\Omega})\}$$

$$(4.11) \qquad L_2^{1,loc}(\Delta,\bar{\Omega}) = L_2^{1,loc}(\bar{\Omega}) \cap L_2^{loc}(\Delta,\bar{\Omega}) .$$

The space $\mathcal{D}'(\Omega)$ in (4.8) is of course the Schwartz space of all distributions on Ω. The notation $\bar{\Omega}$ is used in the definitions to emphasize that the local square-integrability condition are required to hold up to the boundary $\partial\Omega$. The notation $L_2^{loc}(\Omega)$ has sometimes been used for the space $\mathcal{D}'(\Omega) \cap \{u\colon u \in L_2(K)$ for every compact $K \subset \Omega\}$. Functions $u \in L_2^{loc}(\Omega)$ are not restricted near $\partial\Omega$.

The spaces introduced above are all Fréchet spaces (that is, locally convex topological vector spaces which are metrizable and complete [5,30]) under suitable definitions of the topologies. Thus, $L_2^{loc}(\bar{\Omega})$ is a Fréchet space with semi-norms

$$(4.12) \quad \rho_K(u) = \int_{K\cap\Omega} |u(x)|^2 dx, \quad K \text{ a compact subset of } \mathbb{R}^n.$$

Similarly, $L_2^{m,loc}(\bar{\Omega})$ is a Fréchet space with semi-norms

$$(4.13) \quad \begin{cases} \rho_K(u) = \displaystyle\int_{K\cap\Omega} \sum_{|\alpha|\le m} |D^\alpha u(x)|^2 dx, \\ K \text{ a compact subset of } \mathbb{R}^n; \end{cases}$$

$L_2^{loc}(\Delta,\bar{\Omega})$ is a Fréchet space with semi-norms

$$(4.14)\qquad \begin{cases} \rho_K(u) = \displaystyle\int_{K\cap\Omega} \{|u(x)|^2 + |\Delta u(x)|^2\}dx, \\ K \text{ a compact subset of } \mathbb{R}^n; \end{cases}$$

and $L_2^{1,loc}(\Delta,\bar{\Omega})$ is a Fréchet space with semi-norms

$$(4.15)\qquad \begin{cases} \rho_K(u) = \displaystyle\int_{K\cap\Omega} \{|u(x)|^2 + \sum_{k=1}^{n} |D_k u(x)|^2 \\ \qquad + |\Delta u(x)|^2\}dx, \quad K \text{ a compact subset of } \mathbb{R}^n. \end{cases}$$

The following additional notation is used below:

$$(4.16)\qquad L_2^{vox}(\bar{\Omega}) = L_2(\Omega) \cap \mathcal{E}'(\mathbb{R}^n),$$

$$(4.17)\qquad L_2^{1,vox}(\bar{\Omega}) = L_2^1(\Omega) \cap L_2^{vox}(\bar{\Omega}).$$

In (4.16), $\mathcal{E}'(\mathbb{R}^n)$ denotes the Schwartz space of distributions with compact support in $\mathbb{R}^n$.

The generalized Neumann condition of Lecture 3 will be extended to functions in $L_2^{1,loc}(\Delta,\bar{\Omega})$ by the

DEFINITION. A function $u \in L_2^{1,loc}(\Delta,\bar{\Omega})$ is said to satisfy the *generalized Neumann condition* for Ω if and only if

$$(4.18)\qquad \int_\Omega \{(\Delta u)v + \nabla u \cdot \nabla v\}dx = 0 \quad \text{for all} \quad v \in L_2^{1,vox}(\bar{\Omega}).$$

It is easy to verify that conditions (3.10) and (4.18) are equivalent for functions $u \in L_2^1(\Delta,\Omega)$. Moreover, (4.18) is meaningful for arbitrary domains Ω and reduces to the classical Neumann condition whenever $\partial\Omega$ is sufficiently regular.

The solution of the steady-state scattering problem will be constructed in the function space

$$(4.19) \quad L_2^{N,loc}(\Delta,\bar{\Omega}) = L_2^{1,loc}(\Delta,\bar{\Omega}) \cap \{u: u \text{ satisfies } (4.18)\}.$$

It is easy to verify that $L_2^{N,loc}(\Delta,\bar{\Omega})$ is a closed subspace of $L_2^{1,loc}(\Delta,\bar{\Omega})$ and hence is itself a Fréchet space. This has the great advantage that limits of solutions of the steady-state scattering problem will necessarily be solutions.

The notation

$$(4.20) \qquad \Omega_R = \Omega \cap \{x: |x| < R\}$$

is used below. It will be convenient to identify $L_2(\Omega_R)$ with the subspace of $L_2(\Omega)$ consisting of functions with support in Ω_R. Note that $f \in L_2^{vox}(\bar{\Omega})$ if and only if $f \in L_2(\Omega_R)$ for some R.

Solutions of the steady-state scattering problem will be defined and constructed for source functions $f \in L_2^{vox}(\bar{\Omega})$. Suppose that $u \in \mathcal{D}'(\Omega)$ is a solution of

$$(4.21) \qquad \Delta u + \omega^2 u = -f \in L_2(\Omega_R)$$

with $\omega \geq 0$. Then it follows from standard regularity theory for elliptic operators that $u(x)$ is analytic for $|x| > R$ and satisfies

$$(4.22) \qquad \Delta u + \omega^2 u = 0 \quad \text{for all } |x| > R.$$

Hence the Sommerfeld condition in its original form is

applicable to u.

DEFINITION. A solution u of (4.21) is said to satisfy the *Sommerfeld outgoing* (resp., *incoming*) *radiation condition* if and only if

$$(4.23)\quad \begin{cases} \dfrac{\partial u}{\partial |x|} - i\omega u = o\left(\dfrac{1}{|x|^{\frac{n-1}{2}}}\right) \\ \left(\text{resp.,}\quad \dfrac{\partial u}{\partial |x|} + i\omega u = o\left(\dfrac{1}{|x|^{\frac{n-1}{2}}}\right)\right), \end{cases}$$

$$(4.24)\qquad u(x) = O\left(\frac{1}{|x|^{\frac{n-1}{2}}}\right) \text{ for } |x| \to \infty.$$

The limits in (4.23) and (4.24) are understood to be uniform with respect to the direction $x/|x|$.

The class of solutions of the steady-state scattering problem is described by the

DEFINITION. A function $u: \Omega \to \mathbb{C}$ is said to be an *outgoing* (resp., *incoming*) *solution* of the steady-state scattering problem for domain Ω, source function $f \in L_2^{vox}(\bar{\Omega})$ and frequency $\omega \geq 0$ if and only if

$$(4.25)\qquad u \in L_2^{N,loc}(\Delta,\bar{\Omega})$$

$$(4.26)\qquad \Delta u + \omega^2 u = -f \quad \text{in } \Omega$$

and

(4.27) $\begin{cases} \text{u satisfies the outgoing (resp., incoming)} \\ \text{radiation condition.} \end{cases}$

The first property that will be proved for such solutions is their uniqueness. No restrictions on $\partial\Omega$ are needed for this result. Note that (4.25) includes the "edge condition" of diffraction theory [11]. The uniqueness theorem is known to be false if the edge condition is omitted.

THEOREM 4.1. *Let* Ω *be an arbitrary exterior domain. Then the steady-state scattering problem for* Ω, $f \in L_2^{\text{vox}}(\bar{\Omega})$ *and* $\omega \geq 0$ *has at most one solution.*

The proof of this result will be based on a classical theorem of F. Rellich [32] which may be stated as follows.

THEOREM 4.2. (F. Rellich). *Let* $u(x)$ *be a solution of*

$$\Delta u + \omega^2 u = 0 \quad \text{for} \quad |x| > R \tag{4.28}$$

with $\omega^2 > 0$. *Then either* $u(x) \equiv 0$ *for* $|x| > R$ *or for every pair of numbers* R_0, R_1 *such that* $R < R_0 < R_1$ *there exists a constant* $M = M(u, R_0, R_1, \omega) > 0$ *such that*

$$\int_{R_0 \leq |x| \leq r} |u(x)|^2 dx \geq Mr \quad \text{for all} \quad r \geq R_1. \tag{4.29}$$

PROOF OF THEOREM 4.1. To prove the uniqueness of the solution it is enough to prove that if $f(x) = 0$ in Ω then $u(x) = 0$ in Ω because conditions (4.25), (4.26) and (4.27) are linear. The cases $\omega > 0$ and $\omega = 0$ must be treated separately because Rellich's theorem is not valid for $\omega = 0$.

THE CASE $\omega > 0$. The classical proof, applicable when $\partial\Omega$ is smooth, makes use of Green's theorem and the differential equation, boundary condition and radiation condition to show that any solution with $f(x) = 0$ in Ω must satisfy

$$\lim_{r\to\infty} \int_{|x|=r} |u(x)|^2 dS = 0. \tag{4.30}$$

This is inconsistent with (4.29). Hence $u(x) \equiv 0$ for $|x| > R$. But then $u(x) \equiv 0$ in Ω, by the unique continuation property for solutions of $\Delta u + \omega^2 u = 0$ in Ω. Thus to prove theorem 4.1 with $\omega > 0$ it is enough to show that (4.25), (4.26) with $f = 0$ and (4.27) imply (4.30).

To prove (4.30) introduce a function $\phi_1 \in C^\infty(\mathbb{R})$ such that $\phi_1'(\tau) \geq 0$ for $\tau \in \mathbb{R}$, $\phi_1(\tau) \equiv 0$ for $\tau \leq 0$ and $\phi_1(\tau) \equiv 1$ for $\tau \geq 1$. It follows that $0 \leq \phi_1(\tau) \leq 1$ for $\tau \in \mathbb{R}$. Define the function

$$\chi_{r,\delta}(x) = \phi_1\left(\frac{r-|x|}{\delta}\right), \quad x \in \mathbb{R}^n, \ \delta > 0, \ r > 0. \tag{4.31}$$

Then $\chi_{r,\delta}(x) = 1$ for $|x| \leq r - \delta$, $\chi_{r,\delta}(x) = 0$ for $|x| \geq r$ and

$$\begin{cases} \lim_{\delta\to 0} \chi_{r,\delta}(x) = \chi_r(x) = \text{characteristic function} \\ \text{of } \{x: |x| < r\}. \end{cases} \tag{4.32}$$

Moreover, $\chi_{r,\delta} \in C_0^\infty(\mathbb{R}^n)$ for $\delta > 0$. Thus if u is a solution of (4.25), (4.26), (4.27) with $f = 0$ then $v(x) = \chi_{r,\delta}(x)\overline{u}(x) \in L_2^{1,\text{vox}}(\overline{\Omega})$ and hence (4.18) holds with this choice of v. Now the calculus of derivatives in $\mathcal{D}'(\Omega)$ implies

$$\nabla v = \chi_{r,\delta} \nabla\bar{u} + \bar{u}\,\nabla\chi_{r,\delta}. \tag{4.33}$$

Substitution in (4.18) gives

$$\int_\Omega \{(\Delta u)\bar{u}\chi_{r,\delta} + \nabla u \cdot \nabla\bar{u}\chi_{r,\delta} + \bar{u}\nabla u \cdot \nabla\chi_{r,\delta}\}dx = 0. \tag{4.34}$$

Taking the complex conjugate of (4.34) and subtracting it from (4.34) gives

$$\int_\Omega \{u\Delta\bar{u} - \bar{u}\Delta u\}\chi_{r,\delta}dx + \int_\Omega (u\nabla\bar{u} - \bar{u}\nabla u) \cdot \nabla\chi_{r,\delta}dx = 0. \tag{4.35}$$

Moreover, the integral on the left in (4.35) is zero because $\Delta u + \omega^2 u = 0$ in Ω. Also,

$$\nabla\chi_{r,\delta}(x) = \phi_1'\left(\frac{r-|x|}{\delta}\right)\left(-\frac{x}{\delta|x|}\right). \tag{4.36}$$

Introducing spherical coordinates $x = |x|\eta$, $|\eta| = 1$ and (4.36) in (4.35) gives

$$\left\{\int_{r-\delta}^{r} \int_{S^{n-1}} \left\{u\,\frac{\partial\bar{u}}{\partial|x|} - \bar{u}\,\frac{\partial u}{\partial|x|}\right\} \phi_1'\left(\frac{r-|x|}{\delta}\right)|x|^{n-1}d\eta d|x| = 0. \right. \tag{4.37}$$

Now make $\delta \to 0$. The calculation of the limit in (4.37) is classical, since $u \in C^\infty(\Omega)$, and gives

$$\left\{\begin{array}{l} \displaystyle\int_{|x|=r} \left\{u(x)\,\frac{\partial\bar{u}(x)}{\partial|x|} - \bar{u}(x)\,\frac{\partial u(x)}{\partial|x|}\right\} dS = 0 \\ \quad \text{for all} \quad r > R \end{array}\right. \tag{4.38}$$

where $\partial\Omega \subset \Omega_R$. The remainder of the proof is classical. Substituting from (4.23) and (4.24) in (4.38) gives

$$(4.39)\quad \left\{\int_{|x|=r} \left\{ u(x)\,(\mp i\omega \overline{u(x)}) - \overline{u(x)}\,(\pm i\omega u(x)) + o\left(\frac{1}{|x|^{n-1}}\right)\right\} dS = 0.\right.$$

It follows by the uniformity of (4.23), (4.24) in $\eta = x/|x|$ that

$$(4.40)\quad (\pm 2i\omega)\int_{|x|=r} |u(x)|^2 dS = o(1), \quad r \to \infty,$$

which is equivalent to (4.30) when $\omega \neq 0$ and hence completes the proof in this case.

THE CASE $\omega = 0$. In this case u is harmonic in Ω. Hence (4.34) holds with $\Delta u = 0$:

$$(4.41)\quad \int_{\Omega} |\nabla u|^2 \chi_{r,\delta} dx + \int_{\Omega} \overline{u}\nabla u \cdot \nabla\chi_{r,\delta} dx = 0.$$

Moreover, $u(x)$ is analytic in Ω (elliptic regularity) and hence passage to the limit $\delta \to 0$ gives

$$(4.42)\quad \int_{\Omega r} |\nabla u(x)|^2 dx - \int_{|x|=r} \overline{u}\,\frac{\partial u}{\partial |x|}\, dS = 0 \quad \text{for all } r > R.$$

Substituting $\overline{u}$ and $\partial u/\partial|x|$ from (4.23), (4.24) with $\omega = 0$ in the last integral gives

$$(4.43)\quad \int_{|x|=r} \overline{u}\,\frac{\partial u}{\partial |x|}\, dS = \int_{|x|=r} o\left(\frac{1}{|x|^{n-1}}\right) dS = o(1), \quad r \to \infty.$$

Thus making $r \to \infty$ in (4.42) gives

$$\int_\Omega |\nabla u(x)|^2 dx = 0, \tag{4.44}$$

and hence $u(x) = c$, a constant, in Ω. But then (4.24) implies that $c = 0$ which completes the uniqueness proof.

The results presented up to this point are valid for arbitrary exterior domains Ω. Most of the results presented in the remainder of the lectures are based on a local compactness theorem for Ω which is known to be valid only for a restricted class of domains. The theorem states that if $S \subset L_2^{1,loc}(\bar{\Omega})$ is a set of functions whose restrictions to Ω_R are bounded in $L_2^1(\Omega_R)$ then they are precompact in $L_2(\Omega_R)$. This is the famous "selection theorem" of F. Rellich [31]. A more precise formulation will be based on the

DEFINITION. A domain $\Omega \subset \mathbb{R}^n$ is said to have the *local compactness property* if and only if for each set of functions $S \subset L_2^{1,loc}(\bar{\Omega})$ and each $R > 0$ the condition

$$\|u\|_{L_2^1(\Omega_R)} \le C(R) \quad \text{for all} \quad u = v|_{\Omega_R} \quad \text{with} \quad v \in S \tag{4.45}$$

implies that $\{u = v|_{\Omega_R} : v \in S\}$ is precompact in $L_2(\Omega_R)$; i.e., every sequence $\{v_n\}$ in S has a subsequence $\{\tilde{v}_n\}$ such that $u_n = \tilde{v}_n|_{\Omega_R}$ converges in $L_2(\Omega_R)$. The class of domains with the local compactness property will be denoted by LC.

Rellich's original selection theorem stated that

bounded domains with smooth boundaries are in the class LC. The version that will be used here is due, in essence, to S. Agmon who proved the theorem for domains having the "segment property" [1, Theorem 3.8]. This means that there exists a finite open covering of $\partial\Omega$

(4.46) $$\partial\Omega \subset \mathcal{O}_1 \cup \mathcal{O}_2 \cup \dots \cup \mathcal{O}_N$$

and corresponding non-zero vectors $x^{(1)}, x^{(2)}, \dots, x^{(N)}$ such that the segment $\{x = x_0 + tx^{(j)} : 0 < t < 1\} \subset \Omega$ for every $x_0 \in \bar{\Omega} \cap \mathcal{O}_j$. This property defines a rather large class of domains which includes most of the domains that arise in applications. However, there are some unfortunate exceptions. For example, the disk

(4.47) $$\Gamma = \{(x_1, x_2, x_3) : x_1^2 + x_2^2 \le 1 \text{ and } x_3 = 0\} \subset \mathbb{R}^3$$

does not have the segment property. This is because the open sets $\mathcal{O}_j$ necessarily overlap the edge of the disk. The author has observed that the local compactness theorem can be extended to a larger class of domains, while retaining Agmon's method of proof, by modifying the segment property. The first step is to replace the open sets $\mathcal{O}_j$ by compact sets. Assume that there exists an open set $\mathcal{O} \subset \mathbb{R}^n$, compact sets $K_1, K_2, \dots, K_N \subset \mathbb{R}^n$ and non-zero vectors $x^{(1)}, x^{(2)}, \dots, x^{(N)}$ such that

(4.48) $$\partial\Omega \subset \mathcal{O}$$

(4.49) $$\mathcal{O} \cap \Omega \subset \bigcup_{j=1}^{N} K_j$$

(4.50) $\{x = x_0 + tx^{(j)} : 0 < t < 1\} \subset \Omega$ for every $x_0 \in \Omega \cap K_j$.

Then Agmon's proof goes through as before with the sets K_j in place of the sets $\mathcal{O}_j$. The class of domains so defined includes cases like (4.47) where $\partial\Omega$ has sharp edges. Thus, for (4.47) it suffices to choose $N = 2$, $x^{(1)} = (0,0,1)$, $x^{(2)} = (0,0,-1)$, $K_1 = \{(x_1, x_2, x_3): x_1^2 + x_2^2 \leq 2,\ 0 \leq x_3 \leq 1\}$, $K_2 = \{(x_1, x_2, x_3): x_1^2 + x_2^2 \leq 2,\ -1 \leq x_3 \leq 0\}$ and $\mathcal{O} = \{(x_1, x_2, x_3): x_1^2 + x_2^2 < 3/2,\ -1/2 < x_3 < 1/2\}$.

The scope of Agmon's method can be extended further. Note that the conclusion of the local compactness theorem is "intrinsic"; that is, independent of the coordinate systems used to describe Ω. However, condition (4.50) is coordinate-dependent. Condition (4.50) can be generalized to the following condition.

(4.51) Each set K_j lies in a (curvilinear) coordinate patch $\mathcal{O}_j$ such that (4.50) holds when $x = (x_1, \ldots, x_n)$ are the corresponding coordinates.

It is assumed, of course, that the metric tensor in each coordinate patch is bounded and uniformly positive. Then Agmon's proof goes through when the appropriate coordinates are used in each K_j.

The important points about properties (4.48), (4.49) and (4.51) are that, first, a neighborhood of $\partial\Omega$ is completely covered by the sets K_j which need not overlap in more than a set of measure zero and, second, each $\Omega \cap K_j$ is covered by a bundle of segments. The situation is reminiscent of the covering of the "surface" $\partial\Omega$ by a finite set of non-overlapping "tiles", the sets K_j. This suggests the following terminology.

DEFINITION. An exterior domain Ω will be said to have the *finite tiling property* if properties (4.48), (4.49) and (4.51) hold for a suitable open set $\mathcal{O}$, compact sets K_j and coordinates patches $\mathcal{O}_j$.

The following extension of the theorems of Rellich and Agmon is valid.

THEOREM 4.3. *If* Ω *is an exterior domain having the finite tiling property then* $\Omega \in LC$.

The details of the proof may be found in the book of Agmon [1] and are not reproduced here.

The limiting absorption theorem will now be formulated and proved. The following notation will be used. A is the selfadjoint operator in $L_2(\Omega)$ constructed in Lecture 3 and

$$(4.52) \qquad R(z) = (A - z)^{-1}$$

is its resolvent operator. It will be convenient to introduce the two-sheeted Riemann surface for the function $\sqrt{z}$. It will be denoted by $\mathbb{C}_{\frac{1}{2}}$ and

$$(4.53) \qquad \mathbb{C}_{\frac{1}{2}}^{+} = \mathbb{C} \cap \{z: 0 < \arg \sqrt{z} < \pi\}$$

will denote the usual "first sheet". The fact that $\sigma(A) \subset \overline{\mathbb{R}_+}$ implies that $R(z)$ is defined, as a bounded operator on $L_2(\Omega)$, for all $z \in \mathbb{C}_{\frac{1}{2}}^{+}$ and defines an analytic function there. The limiting absorption theorem deals with the boundary values of $R(z)$ on the boundary of $\mathbb{C}^+_{\frac{1}{2}}$ in $\mathbb{C}_{\frac{1}{2}}$.

THEOREM 4.4. *Let* Ω *be an exterior domain such that* $\Omega \in LC$ *and let* $f \in L_2^{\text{vox}}(\overline{\Omega})$. *Then:*

(4.54) $R(z)f \in L_2^{N,\text{loc}}(\Delta,\overline{\Omega})$ *for each* $z \in \mathbb{C}^+_{\frac{1}{2}}$.

(4.55) *The mapping* $T: z \to R(z)f \in L_2^{N,\text{loc}}(\Delta,\overline{\Omega})$ *is continuous on* $\mathbb{C}^+_{\frac{1}{2}}$.

(4.56) T *has a continuous extension to* $\overline{\mathbb{C}^+_{\frac{1}{2}}}$, *the closure of* $\mathbb{C}^+_{\frac{1}{2}}$ *in* $\mathbb{C}_{\frac{1}{2}}$.

Denote points on the "upper" and "lower" edges of the boundary of $\mathbb{C}^+_{\frac{1}{2}}$ *by* $\lambda + i0$ *and* $\lambda - i0$, *respectively, where* $\lambda \geq 0$. *Moreover, let* $u^{\pm}_{\omega} = R(\omega^2 \pm i0)f$, $\omega \geq 0$, *denote the limits whose existence is implied by* (4.56). *Then*

(4.57) $$u^{\pm}_{\omega} = R(\omega^2 \pm i0)f \in L_2^{N,\text{loc}}(\Delta,\overline{\Omega})$$

(4.58) $$(\Delta + \omega^2)u^{\pm}_{\omega} = -f \quad \text{in } \Omega$$

(4.59) u^{+}_{ω} *(resp.* u^{-}_{ω}*) satisfies the outgoing (resp., incoming) radiation condition.*

COROLLARY 4.5. *If* Ω *is an exterior domain and* $\Omega \in LC$ *then the steady-state scattering problem has a unique solution for every* $f \in L_2^{\text{vox}}(\overline{\Omega})$ *and frequency*

$\omega \geq 0$.

PROOF OF COROLLARY 4.5. (4.57)-(4.59) state that u_ω^+ (resp. u_ω^-) satisfy the defining properties (4.25)-(4.26). The uniqueness was proved as Theorem 4.1.

The proof of Theorem 4.4 is based on two lemmas which are of independent interest. The first is

LEMMA 4.6. *Let* Ω *be an exterior domain and* $\Omega \in LC$. *Assume that* $\partial\Omega \subset B(r_0)$ *and let* $I = [a,b] \subset \overline{\mathbb{R}}_+$ $\sigma_0 > 0$, $r > r_0$ *and* $r' > r$. *Then there exists a constant* $M = M(I,\sigma_0, r,r') > 0$ *such that*

$$(4.60) \qquad \|R(\lambda \pm i\sigma)f\|_{L_2^1(\Delta,\Omega_{r'})} \leq M \|f\|_{L_2(\Omega_r)}$$

for all $\lambda \in I$, $0 < \sigma \leq \sigma_0$ *and all* $f \in L_2(\Omega_r)$.

PROOF. The proof is by contradiction. If the lemma is false then there exist sequences $\{\lambda_m\}$ in I, $\{\sigma_m\}$ in $(0,\sigma_0]$ and $\{f_m\}$ in $L_2(\Omega_r)$ such that

$$(4.61) \qquad \|f_m\|_{L_2(\Omega_r)} = 1, \quad m = 1, 2, 3, \ldots$$

and

$$(4.62) \quad \|R(\lambda_m \pm i\sigma_m)f_m\|_{L_2^1(\Delta,\Omega_{r'})} > m, \quad m = 1, 2, 3, \ldots .$$

It follows that there exists a subsequence of $\{\lambda_m \pm i\sigma_m\}$ which converges. Denote the subsequence by the same symbol and write $\lim (\lambda_m \pm i\sigma_m) = \lambda \pm i\sigma$. Then σ must be zero, because $R(z)$ is analytic on $C_{\frac{1}{2}}^+$ in the uniform operator topology [12]. Hence if $\sigma \neq 0$ then $\|R(\lambda_m \pm i\sigma_m)\|$ would have a limit when $m \to \infty$ which contradicts (4.62). Thus

$$\lim_{m\to\infty} (\lambda_m \pm i\sigma_m) = \lambda \in I. \tag{4.63}$$

Now define

$$u_m = \frac{R(\lambda_m \pm i\sigma_m) f_m}{\|R(\lambda_m \pm i\sigma_m) f_m\|_{L_2^1(\Delta,\Omega_{r'})}} \tag{4.64}$$

and

$$F_m = \frac{f_m}{\|R(\lambda_m \pm i\sigma_m) f_m\|_{L_2^1(\Delta,\Omega_{r'})}} . \tag{4.65}$$

Note that the denominators in (4.64), (4.65) are not zero by (4.62). Moreover,

$$\begin{cases} \|u_m\|^2_{L_2^1(\Delta,\Omega_{r'})} = \|u_m\|^2_{L_2(\Omega_{r'})} + \sum_{j=1}^{n} \|D_j u_m\|^2_{L_2(\Omega_{r'})} \\ \quad + \|\Delta u_m\|^2_{L_2(\Omega_{r'})} = 1, \end{cases} \tag{4.66}$$

$$\|F_m\|_{L_2(\Omega_r)} < \frac{1}{m}, \tag{4.67}$$

and

$$(\Delta + (\lambda_m \pm i\sigma_m)) u_m = -F_m \quad \text{in } \Omega \tag{4.68}$$

for $m = 1, 2, 3, \ldots$. Now the set $S = \{u_1, u_2, u_3, \ldots\} \subset L_2^N(\Delta,\Omega) \subset L_2^{1,loc}(\bar{\Omega})$ and satisfies

$$(4.69) \qquad \|u_m\|_{L_2^1(\Omega_{r'})} \le 1 \quad \text{for} \quad m = 1, 2, \ldots$$

by (4.66). Moreover, it is assumed that $\Omega \in LC$. Thus Theorem 4.3 is applicable and implies that $\{u_m\}$ has a subsequence which converges in $L_2(\Omega_{r'})$. Denote it again by the same symbol. Then it will be shown that

$\{u_m\}$ CONVERGES IN $L_2^{N,loc}(\Delta,\bar{\Omega})$. To prove this choose a radius r'' such that $r < r'' < r'$. The proof will consist of two steps. Step 1 is the proof that $\{u_m\}$ converges in $L_2^1(\Delta,\Omega_{r''})$. Step 2 is the proof that $\{u_m\}$ converges in $L_2^1(\Delta, \Omega_R - \Omega_{r''})$ for any $R > r''$.

STEP 1. Note that $\operatorname{supp} F_m \subset \Omega_r \subset \Omega_{r'}$ and hence $\lim_{m\to\infty} F_m = 0$ in $L_2(\Omega_{r'})$ by (4.67). It follows by (4.63), (4.68) and the convergence of $\{u_m\}$ in $L_2(\Omega_{r'})$ that $\{\Delta u_m\}$ converges in $L_2(\Omega_{r'})$. To complete the proof of Step 1 it will be shown that $\nabla u_m = (D_1 u_m, \ldots, D_n u_m)$ converges in $L_2(\Omega_{r''})$. Define

$$(4.70) \qquad \psi(x) = \chi_{r',r'-r''}(x), \quad x \in \mathbb{R}^n,$$

where $\chi_{r,\delta}$ is defined by (4.31). ($\psi(x) \equiv 1$ for $|x| \le r''$ and $\psi(x) \equiv 0$ for $|x| \ge r'$.) Now apply the generalized Neumann condition (4.18) to $u = u_\ell - u_m \equiv u_{\ell m}$ and $v = \psi\bar{u}_{\ell m}$. The result is (cf. (4.34)):

$$(4.71) \qquad \int_{\Omega_{r'}} \{(\Delta u_{\ell m})\bar{u}_{\ell m}\psi + |\nabla u_{\ell m}|^2\psi + \bar{u}_{\ell m}\nabla u_{\ell m} \cdot \nabla\psi\}dx = 0.$$

This implies the following estimate for $\nabla u_{\ell m}$:

$$(4.72)\begin{cases} \int_{\Omega_{r''}} |\nabla u_{\ell m}|^2 dx \leq \int_{\Omega_{r'}} |\nabla u_{\ell m}|^2 \psi dx \\ \leq -\int_{\Omega_{r'}} \{(\Delta u_{\ell m})\overline{u}_{\ell m}\psi + \overline{u}_{\ell m}\nabla u_{\ell m}\cdot\nabla\psi\}dx \\ \leq M_1\{\|u_{\ell m}\|_{r'}\|\Delta u_{\ell m}\|_{r'} + \|u_{\ell m}\|_{r'}\|\nabla u_{\ell m}\|_{r'}\} \\ \leq M_1\{\|\Delta u_{\ell m}\|_{r'}^2 + \|u_{\ell m}\|_{r'}^2 + \delta\|\nabla u_{\ell m}\|_{r'}^2 + \frac{1}{\delta}\|u_{\ell m}\|_{r'}^2\} \end{cases}$$

where $M_1 \geq 1$ is a bound for $|\nabla\psi|$, $\delta > 0$ is arbitrary and $\|u\|_{r'}$ has been written for $\|u\|_{L_2(\Omega_{r'})}$. Choose $\delta = \epsilon/M_1$ and write $M_\epsilon = M_1(1 + 1/\delta) = M_1 + M_1^2/\epsilon$. Then (4.72) becomes

$$(4.73)\begin{cases} \|\nabla u_\ell - \nabla u_m\|_{r''}^2 \\ \leq \epsilon\|\nabla u_\ell - \nabla u_m\|_{r'}^2 + M_\epsilon\{\|\Delta u_\ell - \Delta u_m\|_{r'}^2 + \|u_\ell - u_m\|_{r'}^2\}. \end{cases}$$

Now $\|\nabla u_m\|_{r'} \leq 1$ for $m = 1,2,\ldots$ by (4.66). Thus (4.73) implies

$$(4.74)\begin{cases} \|\nabla u_\ell - \nabla u_m\|_{r''}^2 \\ \leq 4\epsilon + M_\epsilon\{\|\Delta u_\ell - \Delta u_m\|_{r'}^2 + \|u_\ell - u_m\|_{r'}^2\}. \end{cases}$$

Making $\ell,m \to \infty$ in (4.74) gives

(4.75) $$\overline{\lim_{\ell,m\to\infty}} \|\nabla u_\ell - \nabla u_m\|^2_{r''} \leq 4\epsilon$$

because $\{u_m\}$ and $\{\Delta u_m\}$ converge in $L_2(\Omega_{r'})$. This proves that $\{\nabla u_m\}$ converges in $L_2(\Omega_{r''})$ since ϵ is arbitrary. This completes Step 1.

STEP 2. The Green's function, or resolvent kernel, for the operator A_0 of Lecture 2 is used in this step. It is given by

(4.76) $$G_0(x,x',z) = \frac{i}{4}\left(\frac{\sqrt{z}}{2\pi R}\right)^{\frac{n-2}{2}} H^{(1)}_{\frac{n-2}{2}}(\sqrt{z}R), \quad R = |x - x'|$$

where $z \in \mathbb{C}^+_{\frac{1}{2}}$ and $H^{(1)}_p$ is the Hankel function of the first kind. It has the property that $(A_0 - z)u = f$ in $L_2(\mathbb{R}^n)$ if and only if

(4.77) $$u(x) = \int_{\mathbb{R}^n} G_0(x, x', z) f(x') dx'.$$

In particular, this implies that as a distribution G_0 satisfies

(4.78) $$(\Delta_x + z) G_0(x, x', z) = -\delta(x - x').$$

Note that each $u_m(x) \in C^\infty(\Omega - \Omega_r)$ because (4.68) holds with $\operatorname{supp} F_m \subset \Omega_r$. Choose a radius r''' such that $r < r''' < r''$ and apply Green's theorem to $u_m(x')$ and $v_m(x') = G_0(x,x',\lambda_m \pm i\sigma_m)$ in the region $\Omega_R - \Omega_{r'''}$ with $r'' \leq |x| < R$. If $S_r = \{x: |x| = r\}$ the result may be written

$$(4.79)\quad \begin{cases} u_m(x) = \displaystyle\int_{S_{r'''}} \left\{ u_m(x') \frac{\partial G_0(x,x',\lambda_m \pm i\sigma_m)}{\partial |x'|} \right. \\ \qquad\qquad \left. - \dfrac{\partial u_m(x')}{\partial |x'|} G_0(x,x',\lambda_m \pm i\sigma_m) \right\} dS' \\ \quad - \displaystyle\int_{S_R} \left\{ u_m(x') \frac{\partial G_0(x,x',\lambda_m \pm i\sigma_m)}{\partial |x'|} \right. \\ \qquad\qquad \left. - \dfrac{\partial u_m(x')}{\partial |x'|} G_0(x,x',\lambda_m \pm i\sigma_m) \right\} dS' \\ \quad = u_m^{(1)}(x) + u_m^{(2)}(x) \end{cases}$$

where $u_m^{(1)}$ and $u_m^{(2)}$ represent the integrals over $S_{r'''}$ and S_R, respectively. Note that $u_m^{(2)}(x)$ is indeed independent of R, since $u_m(x)$ and $u_m^{(1)}(x)$ are. Thus it has an analytic continuation to all points $x \in \mathbb{R}^n$ and satisfies

$$(4.80)\quad (\Delta + (\lambda_m \pm i\sigma_m)^2)u_m^{(2)}(x) = 0 \quad \text{for all} \quad x \in \mathbb{R}^n$$

by (4.78). Recall that $u_m \in L_2(\Omega)$, by (4.64). Moreover, $u_m^{(1)}(x) \in L_2(\Omega - \Omega_{r''})$ because $G_0(x,x',\lambda_m \pm i\sigma_m)$ tends to zero exponentially when $|x'| = r'''$ and $|x| \to \infty$. It follows that $u_m^{(2)} = u_m - u_m^{(1)} \in L_2(\mathbb{R}^n)$. But this is possible only if $u_m^{(2)}(x) \equiv 0$ because A_0 has no eigenvectors in $L_2(\mathbb{R}^n)$. Thus (4.79) implies that $u_m = u_m^{(1)}$ or

$$(4.81)\quad \begin{cases} u_m(x) = \displaystyle\int_{S_{r'''}} \left\{ u_m(x') \frac{\partial G_0(x,x',\lambda_m \pm i\sigma_m)}{\partial |x'|} \right. \\ \qquad \left. - \dfrac{\partial u_m(x')}{\partial |x'|} G_0(x,x',\lambda_m \pm i\sigma_m) \right\} dS' \quad \text{for all } |x| \geq r''. \end{cases}$$

Next, note that $\{u_m\}$ and $\{\nabla u_m\}$ converge in $L_2(S_{r'''})$. To see this note that since $\{u_m\}$ and $\{\Delta u_m\}$ converge in $\Omega_{r'} - \Omega_r$ the standard interior estimates of elliptic theory imply that $\{D^\alpha u_m\}$ converges in $L_2(\Omega_{r'-\delta} - \Omega_{r+\delta})$ for $|\alpha| \leq 2$ [1, Ch. 6]. Choose δ so that the radii are arranged as follows:

$$(4.82)\quad r_0 < r < r+\delta < r''' < r'' < r'-\delta < r' < R.$$

Then Sobolev's embedding theorem [1,22] implies that $\{u_m\}$ and $\{\nabla u_m\}$ converge in $L_2(S_{r'''})$. Note that the integral in (4.81) can be differentiated under the integral sign for any $|x| \geq r''$. Thus

$$(4.83)\quad \begin{cases} D^\alpha u_m(x) = \displaystyle\int_{S_{r'''}} \left\{ u_m(x') \frac{\partial D^\alpha G_0(x,x',\lambda_m \pm i\sigma_m)}{\partial |x'|} \right. \\ \qquad \left. - \dfrac{\partial u_m(x')}{\partial |x'|} D^\alpha G_0(x, x', \lambda_m \pm i\sigma_m) \right\} dS' \end{cases}$$

for all multi-indices α.

The representation (4.83) implies that each $D^\alpha u_m(x)$ converges uniformly on compact subsets of $\Omega - \Omega_{r''}$. This follows easily from the convergence of $\{u_m\}$ and $\{\nabla u_m\}$ in $L_2(S_{r'''})$ and the fact that

$D^{\alpha}G_0(x,x',\lambda_m \pm i\sigma_m)$ and its normal derivative converge uniformly for $x \in \Omega_R - \Omega_{r''}$ and $x' \in S_{r'''}$. In particular, $\{u_m\}$ converges in $L_2^1(\Delta,\Omega_R - \Omega_{r''})$ which completes Step 2.

The proof that $\{u_m\}$ converges in $L_2^{N,loc}(\Delta,\bar{\Omega})$ can now be completed. By Steps 1 and 2, $\{u_m\}$ converges in both $L_2^1(\Delta,\Omega_{r''})$ and $L_2^1(\Delta,\Omega_R - \Omega_{r''})$ for any $R > r''$. Thus, $\{u_m\}$ converges in $L_2^1(\Delta,\Omega_R)$ for all $R > r''$ which implies the convergence in $L_2^{1,loc}(\Delta,\bar{\Omega})$. Finally, each $u_m \in L_2^{N,loc}(\Delta,\bar{\Omega})$ which is a closed subspace of $L_2^{1,loc}(\Delta,\bar{\Omega})$.

THE CONTRADICTION. To complete the proof of Lemma 4.6 it must be shown that the convergence of $\{u_m\}$ in $L_2^{N,loc}(\Delta,\bar{\Omega})$ leads to a contradiction. To show this let

$$u = \lim_{m\to\infty} u_m \in L_2^{N,loc}(\Delta,\bar{\Omega}). \tag{4.84}$$

Then (4.68) implies

$$(\Delta + \lambda)u = 0 \quad \text{in} \quad \Omega \tag{4.85}$$

since $F_m \to 0$ in $L_2^{loc}(\bar{\Omega})$. Moreover, passage to the limit in (4.81) gives for all $|x| \geq r''$,

$$\left\{ \begin{aligned} u(x) = \int_{S_{r'''}} \Big\{ & u(x') \frac{\partial G_0(x,x',\lambda \pm i0)}{\partial |x'|} \\ & - \frac{\partial u(x')}{\partial |x'|} G_0(x,x',\lambda \pm i0) \Big\} dS' \end{aligned} \right. \tag{4.86}$$

which implies the $u(x)$ satisfies the outgoing (+) or

incoming (-) radiation condition. It follows by the uniqueness theorem (Theorem 4.1) that $u(x) \equiv 0$ in Ω. On the other hand, passage to the limit in (4.66) gives

$$(4.87) \qquad \|u\|_{L_2^1(\Delta,\Omega_{r'})} = 1$$

which is clearly a contradiction.

The second lemma needed for the proof of the limiting absorption theorem is

LEMMA 4.7. *Let* Ω *be an exterior domain such that* $\Omega \in LC$. *Let* $I = [a,b] \subset \overline{\mathbb{R}}_+$, $\sigma_0 > 0$ *and* $f \in L_2(\Omega_r)$. *Then the mapping* $T: z \to R(z)f \in L_2^{N,loc}(\Delta,\overline{\Omega})$ *is uniformly continuous on the set* $\{z = \lambda \pm i\sigma: \lambda \in I,\ 0 < \sigma \leq \sigma_0\}$.

PROOF. The proof is again by contradiction. Note first that it may be assumed that $\partial\Omega \subset B(r)$ since $r < r'$ implies $L_2(\Omega_r) \subset L_2(\Omega_{r'})$. Now if the lemma is false then there must exist an $r' > r$, an $\epsilon > 0$ and sequences $\{\lambda_m\}$, $\{\nu_m\}$ in I and $\{\sigma_m\}$, $\{\tau_m\}$ in $(0,\sigma_0]$ such that

$$(4.88) \qquad |\lambda_m - \nu_m| < \frac{1}{m} \quad \text{and} \quad |\sigma_m - \tau_m| < \frac{1}{m} \quad \text{for } m = 1,2,\ldots$$

and

$$(4.89) \qquad \begin{cases} \|R(\lambda_m \pm i\sigma_m)f - R(\nu_m \pm i\tau_m)f\|_{L_2^1(\Delta,\Omega_{r'})} \geq \epsilon \\ \qquad \text{for } m = 1,2,\ldots\ . \end{cases}$$

It follows that there exist subsequences, to be denoted by the same symbols, such that

$$\zeta_m = \lambda_m \pm i\sigma_m \to \lambda, \quad \mu_m = \nu_m \pm i\tau_m \to \lambda \quad \text{when} \quad m \to \infty, \tag{4.90}$$

where $\lambda \in I$. The two limits must be the same because of (4.88). The limit is real because $u_z = R(z)f$ is continuous in $L_2^1(\Delta,\Omega_{r'})$ at all non-real points. The proof of this statement is based on the continuity of u_z and $\Delta u_z = -zu_z - f$ in $L_2(\Omega)$. The continuity of ∇u_z can be proved by the argument used in the proof of Lemma 4.6, Step 1, to prove the convergence of $\{\nabla u_m\}$.

Now Lemma 4.6 implies that there exists a constant $M = M(I,\sigma_0,r',r)$ such that

$$\left\{\begin{aligned} &\|R(\zeta_m)f\|_{L_2^1(\Delta,\Omega_{r'})} \le M\,\|f\|_{L_2(\Omega_r)} \\ &\|R(\mu_m)f\|_{L_2^1(\Delta,\Omega_{r'})} \le M\,\|f\|_{L_2(\Omega_r)} \end{aligned}\right. \tag{4.91}$$

for $m = 1, 2, 3, \ldots$. Thus the local compactness theorem is applicable to the sequences $\{R(\zeta_m)f\}$ and $\{R(\mu_m)f\}$. Hence there exist subsequences, denoted again by $\{\zeta_m\}$, $\{\mu_m\}$, such that the limits

$$\left\{\begin{aligned} &\lim_{m\to\infty} R(\zeta_m)f = u_\lambda \quad \text{and} \quad \lim_{m\to\infty} R(\mu_m)f = v_\lambda \\ &\text{exist in } L_2^{N,loc}(\Delta,\overline{\Omega}). \end{aligned}\right. \tag{4.92}$$

Moreover, the limits must satisfy

$$(\Delta + \lambda)u_\lambda = -f \quad \text{and} \quad (\Delta + \lambda)v_\lambda = -f \quad \text{in } \Omega \tag{4.93}$$

and the outgoing (+) or incoming (-) radiation condition, since each will have a representation of the form (4.86). But then $u_\lambda = v_\lambda$ in Ω by Theorem 4.1, and hence $\lim_{m\to\infty} R(\zeta_m)f = u_\lambda = \lim_{m\to\infty} R(\mu_m)f$ in $L_2^{N,loc}(\Delta,\overline{\Omega})$ which contradicts (4.89).

PROOF OF THEOREM 4.4. To prove (4.54) note that $\mathbb{C}_{\frac{1}{2}}^{+}$ is contained in the resolvent set of A. To prove (4.55) note that $u_z = R(z)f$ and $\Delta u_z = -zu_z - f$ are continuous in $L_2(\Omega)$ for $z \in \mathbb{C}_{\frac{1}{2}}^{+}$. The proof that ∇u_z is continuous in $L_2(\Omega)$ follows from the generalized Neumann condition (3.10) with $u = v = u_z - u_{z'}$. Thus $R(z)f$ is continuous in $L_2^N(\Delta,\Omega) \subset L_2^{N,loc}(\Delta,\overline{\Omega})$ for all $z \in \mathbb{C}_{\frac{1}{2}}^{+}$. The proof of (4.56), the existence of a continuous extension of T to $\overline{\mathbb{C}_{\frac{1}{2}}^{+}}$ is an immediate consequence of Lemma 4.7 and the completeness of $L_2^{N,loc}(\Delta,\overline{\Omega})$. Conclusions (4.57) and (4.58) also follow from the completeness. Finally, (4.59) follows because $u_\omega^{\pm}$ has a representation of the form (4.86).

LECTURE 5. TIME-DEPENDENT SCATTERING THEORY IN EXTERIOR DOMAINS

In this lecture the time-dependent scattering theory is developed for the pair of operators $A^{\frac{1}{2}}$, $A_0^{\frac{1}{2}}$. The wave operators for the pair are defined and their basic properties are formulated. The relationship of these wave operators to scattering theory for the d'Alembert equation in Ω is discussed at the end of this lecture and in Lectures 7 and 8. The lecture begins with some preliminary results on the spectra of A_0 and A and on the asymptotic behavior of wave functions in compact sets.

The form of the spectral family $\{\Pi_0(\mu)\}$ for A_0, given by (2.24), implies that A_0 is spectrally absolutely continuous [18, Ch. X]. Indeed, (2.24) implies that

$$
(5.1) \qquad \begin{cases} \|\Pi_0(\mu) f\|^2 = \displaystyle\int_{|p|\le\sqrt{\mu}} |\hat{f}(p)|^2 dp \\ \text{for all } \mu \geq 0 \text{ and } f \in L_2(\mathbb{R}^n). \end{cases}
$$

Moreover, a short calculation gives

$$
(5.2) \qquad \frac{d}{d\mu}\|\Pi_0(\mu) f\|^2 = \tfrac{1}{2}\mu^{\frac{n-2}{2}} \int_{S^{n-1}} |\hat{f}(\sqrt{\mu}\eta)|^2 d\eta, \quad \mu > 0.
$$

It is clear that for any $\mu > 0$ this quantity will be positive for suitable choices of f. It follows that the spectrum of A_0 is $\overline{\mathbb{R}}_+ = \{\lambda: \lambda \geq 0\}$: $\sigma(A_0) = \overline{\mathbb{R}}_+$. With regard to A, the positivity $A \geq 0$ implies that $\sigma(A) \subset \overline{\mathbb{R}}_+$. The equality of $\sigma(A) = \overline{\mathbb{R}}_+$ is proved at the end of this lecture.

The theorem of F. Rellich on solutions of $\Delta u + \omega^2 u = 0$ in exterior domains, quoted above as Theorem 4.2, implies

THEOREM 5.1. *The selfadjoint operator* A *corresponding to an arbitrary exterior domain has no eigenvalues.*

PROOF. Suppose that $u \in D(A)$ and $Au = \lambda u$ with $\lambda \in \overline{\mathbb{R}}_+$. It will be shown that u is necessarily zero. First, suppose that $\lambda = \omega^2 > 0$. Then Rellich's theorem is applicable and if $u(x) \not\equiv 0$ for $|x| > R$ then (4.29) holds. But this is impossible with $u \in L_2(\Omega)$. Thus $u(x) \equiv 0$ for all $|x| > R$. This implies that $u = 0$ in $L_2(\Omega)$, since $u(x)$ is necessarily analytic in Ω. Finally, if $\lambda = 0$ then $u \in L_2^N(\Delta,\Omega)$ and $\Delta u = 0$ in Ω. Thus (3.10) holds with $v = u$ and $\Delta u = 0$; that is, $\|\nabla u\| = 0$. This implies that $u(x) \equiv c$, a constant, and hence $u(x) \equiv 0$, since $u \in L_2(\Omega)$.

Theorem 5.1 implies the continuity of the spectral family $\{\Pi(\mu)\}$ for A:

(5.3) $\|\Pi(\mu) f\|^2 = (\Pi(\mu) f, f) \in C(\mathbb{R})$ for all $f \in L_2(\Omega)$.

A proof may be found in the book of T. Kato [18, Ch. X]. Now write $\Pi(I) = \Pi(b) - \Pi(a)$, where $I = (a,b)$, and define

(5.4) $m_f(I) = \|\Pi(I) f\|^2 = (\Pi(I) f, f)$, $f \in L_2(\Omega)$.

Then $m_f(I)$ defines a measure on the ring of intervals I and hence has a unique extension to a measure $m_f(S)$ on the σ-ring of Borel subsets of $\mathbb{R}$. It is known that [10,18]

$$(5.5) \qquad m_f(S) = \|\Pi(S)f\|^2 \quad \text{for all Borel sets } S$$

where $\Pi(S)$ is an orthogonal projection in $L_2(\Omega)$. Define the subsets

$$(5.6) \qquad \mathcal{H}^{ac}(A) = L_2(\Omega) \cap \{f: m_f(S) \text{ is absolutely continuous}\}$$

$$(5.7) \qquad \mathcal{H}^{sc}(A) = L_2(\Omega) \cap \{f: m_f(S) \text{ is singularly continuous}\}$$

where absolute and singular continuity refer to the Lebesgue measure on $\mathbb{R}$. Then Kato has proved [18, Ch. X]

THEOREM 5.2. $\mathcal{H}^{ac}(A)$ *and* $\mathcal{H}^{sc}(A)$ *are closed subspaces of* $L_2(\Omega)$ *which are orthogonal and reduce the operator* A. *Moreover,*

$$(5.8) \qquad L_2(\Omega) = \mathcal{H}^{ac}(A) \oplus \mathcal{H}^{sc}(A).$$

Kato has called $\mathcal{H}^{ac}(A)$ and $\mathcal{H}^{sc}(A)$ the subspaces of absolute continuity and singular continuity for A, respectively. It will be shown next that for exterior domains $\Omega \in LC$ there is no singular continuous spectrum: $\mathcal{H}^{sc}(A) = \{0\}$.

THEOREM 5.3. *Let* Ω *be an exterior domain such that* $\Omega \in LC$. *Then*

$$(5.9) \qquad L_2(\Omega) = \mathcal{H}^{ac}(A),$$

that is, $m_f(S)$ *is absolutely continuous for all* $f \in L_2(\Omega)$.

The proof of Theorem 5.3 is based on the limiting absorption theorem of Lecture 4 and the following well-known theorem of M. H. Stone [6,18,38].

THEOREM 5.4. *Let* $\mathcal{H}$ *be an abstract Hilbert space and* H *a selfadjoint operator on* $\mathcal{H}$ *with spectral family* $\{\Pi(\lambda)\}$ *and resolvent* $R(z) = (H - z)^{-1}$. *Let* (a,b) *be any finite interval. Then*

$$(5.10) \quad \begin{cases} (f, [\Pi(b) + \Pi(b-) - \Pi(a) - \Pi(a-)]g) \\ \quad = \lim_{\sigma \to 0+} \frac{1}{\pi i} \int_a^b (f, [R(\lambda + i\sigma) - R(\lambda - i\sigma)]g)\, d\lambda. \end{cases}$$

PROOF OF THEOREM 5.3. To begin let $I = (a,b)$ be any interval and let $f \in L_2^{vox}(\bar{\Omega})$. Apply (5.10) with $g = f$. The result can be written, by (5.3), as

$$(5.11) \quad \begin{cases} m_f(I) = (f, \Pi(I) f) \\ \quad = \lim_{\sigma \to 0+} \frac{1}{2\pi i} \int_I (f, [R(\lambda + i\sigma) - R(\lambda - i\sigma)] f)\, d\lambda. \end{cases}$$

Now the integrand in this integral is a continuous function of $(\lambda,\sigma) \in I \times [0,\sigma_0]$ by Theorem 4.4. (Recall that supp f is compact.) Thus it is uniformly continuous (cf. Lemma 4.7) and (5.11) implies that

$$(5.12) \quad m_f(I) = \frac{1}{2\pi i} \int_I (f, [R(\lambda + i0) - R(\lambda - i0)] f)\, d\lambda$$

where the integrand is continuous for all $f \in L_2^{vox}(\bar{\Omega})$ and $\lambda \in \mathbb{R}$. It follows that the corresponding Borel extension of $m_f(I)$ satisfies

$$(5.13)\quad m_f(S) = \frac{1}{2\pi i}\int_S (f,[R(\lambda+i0)-R(\lambda-i0)]f)\,d\lambda,\quad f \in L_2^{vox}(\bar{\Omega}).$$

In particular, if $|S|$ denotes the Borel measure of S then $|S| = 0$ implies $m_f(S) = 0$ for all $f \in L_2^{vox}(\bar{\Omega})$. To extend this result to arbitrary $f \in L_2(\Omega)$ recall that $\mathcal{H}^{ac}(A)$ is a closed subspace (Theorem 5.2). Moreover, $L_2^{vox}(\bar{\Omega})$ is dense in $L_2(\Omega)$ and it has just been proved that $L_2^{vox}(\bar{\Omega}) \subset \mathcal{H}^{ac}(A)$. Equation (5.9) follows immediately.

Now consider the complex-valued solutions of the d'Alembert equation which were introduced in Lecture 3:

$$(5.16)\qquad v(t,\cdot) = e^{-itA^{\frac{1}{2}}}h,\quad h \in L_2(\Omega).$$

It will be shown that Theorem 5.3 implies

THEOREM 5.5. *Let* Ω *be an exterior domain such that* $\Omega \in LC$. *Then* $v(t,\cdot)$ *tends to zero in* $L_2^{loc}(\bar{\Omega})$ *for each* $h \in L_2(\Omega)$; *that is,*

$$(5.17)\quad \lim_{t\to\infty}\int_{K\cap\Omega}|v(t,x)|^2dx = 0 \textit{ for each bounded } K \subset \mathbb{R}^n.$$

It is known that this kind of "local decay" property is closely related to the question of the existence of wave operators; see [44] for a discussion. Theorem 5.5 is used in Lecture 7 in the construction of the wave operators for $A^{\frac{1}{2}}$ and $A_0^{\frac{1}{2}}$ given there. The analogue of (5.17) for the energy, mentioned in Lecture 1, is proved in Lecture 8.

PROOF OF THEOREM 5.5. Let $Q_K: L_2(\Omega) \to L_2(\Omega)$ denote the orthogonal projection defined by $Q_K u(x) = \chi_K(x)u(x)$ where $\chi_K(x)$ is the characteristic function of K. Then (5.17) is equivalent to

$$(5.18) \quad \lim_{t\to\infty} \|Q_K e^{-itA^{\frac{1}{2}}} h\| = 0 \quad \text{for each bounded } K \subset \mathbb{R}^n.$$

The proof of (5.18) will be based on an abstract theorem which is given in [44]. It states that (5.18) holds for every $h \in \mathcal{H}^{ac}(A^{\frac{1}{2}})$ if Q_K is $A^{\frac{1}{2}}$-compact [18] for each bounded K. Note that Theorem 5.3 implies $L_2(\Omega) = \mathcal{H}^{ac}(A^{\frac{1}{2}})$, by the spectral theorem. Thus, the proof will be complete when Q_K has been shown to be $A^{\frac{1}{2}}$-compact. This means that any set $S \subset L_2(\Omega)$ which is bounded in the graph norm of $A^{\frac{1}{2}}$ must have the property that $Q_K S$ is precompact in $L_2(\Omega)$. Now $D(A^{\frac{1}{2}}) = L_2^1(\Omega)$ and the graph norm of $A^{\frac{1}{2}}$ is $\{\|A^{\frac{1}{2}}u\|^2 + \|u\|^2\}^{\frac{1}{2}} = \{\|\nabla u\|^2 + \|u\|^2\}^{\frac{1}{2}} = \|u\|_1$. Thus the $A^{\frac{1}{2}}$-compactness of Q_K follows from Theorem 4.3.

Theorem 5.5 states that solutions in $L_2(\Omega)$ of the d'Alembert equation tend to zero in any bounded neighborhood of $\Gamma = \mathbb{R}^n - \Omega$. This suggests that each wave (5.16) should be asymptotically equal to a free wave $v_0(t,x)$ in $\mathbb{R}^n$; that is,

$$(5.19) \qquad \lim_{t\to\infty} \left\| e^{-itA^{\frac{1}{2}}} h - e^{-itA_0^{\frac{1}{2}}} h_0 \right\|_{L_2(\Omega)} = 0$$

where h_0 is a suitable function in $L_2(\mathbb{R}^n)$. Note that, since (5.18) holds for A_0, (5.19) is equivalent to the condition

$$(5.20)\qquad \lim_{t\to\infty} \| J_\Omega e^{-itA^{\frac{1}{2}}} h - e^{-itA_0^{\frac{1}{2}}} h_0 \|_{L_2(\mathbb{R}^n)} = 0$$

where $J_\Omega : L_2(\Omega) \to L_2(\mathbb{R}^n)$ is defined by

$$(5.21)\qquad J_\Omega u(x) = \begin{cases} u(x), & x \in \Omega, \\ 0, & x \in \mathbb{R}^n - \Omega. \end{cases}$$

Now (5.20) is equivalent to

$$(5.22)\qquad \lim_{t\to\infty} \| e^{itA_0^{\frac{1}{2}}} J_\Omega e^{-itA^{\frac{1}{2}}} h - h_0 \|_{L_2(\mathbb{R}^n)} = 0$$

because $e^{-itA_0^{\frac{1}{2}}}$ is a unitary operator on $L_2(\mathbb{R}^n)$. Finally, (5.22) can be formulated without reference to the unknown function h_0 as the statement that the strong operator limit

$$(5.23)\qquad W_+ = W_+(A_0^{\frac{1}{2}}, A^{\frac{1}{2}}, J_\Omega) = \operatorname*{s-lim}_{t\to+\infty} e^{itA_0^{\frac{1}{2}}} J_\Omega e^{-itA^{\frac{1}{2}}}$$

should exist. The existence of the operator W_+ is entirely equivalent to the asymptotic condition (5.19) with $h_0 = W_+ h$. W_+ is called the wave operator for the triple $A_0^{\frac{1}{2}}$, $A^{\frac{1}{2}}$, J_Ω. The construction of the wave operator is one of the primary goals of the time-dependent theory of scattering. A construction will be given in Lecture 7, on the basis of the eigenfunction expansion for A which is developed in Lecture 6. One consequence of the construction will be stated here as

THEOREM 5.6. *If* Ω *is an exterior domain such that* $\Omega \in LC$ *then the wave operator*

$$W_+ : L_2(\Omega) \to L_2(\mathbb{R}^n) \tag{5.24}$$

exists and is unitary. Moreover, it maps the spectral family of A_0 *into that of* A; *that is,*

$$\Pi(\lambda) = W_+^* \Pi_0(\lambda) W_+ \quad \textit{for all} \quad \lambda \in \mathbb{R} \tag{5.25}$$

where $W_+^* = W_+^{-1}$.

COROLLARY 5.7. *The operators* A *and* A_0 *are unitarily equivalent. In particular,*

$$\sigma(A) = \sigma(A_0) = \overline{\mathbb{R}}_+ . \tag{5.26}$$

Theorem 5.6 follows from Corollary 7.2 of Lecture 7. Another proof of it can be given on the basis of an abstract operator-theoretic existence theorem for wave operators due to A. L. Belopolskii and M. S. Birman [2]. A complete proof by this method is outside the scope of these lectures because the Belopolskii-Birman theory is based on a long theoretical development. However, because of the great methodological interest of the Belopolskii-Birman theory, it is described briefly, and applied to the proof of Theorem 5.6 in an Appendix to these lectures.

LECTURE 6. STEADY-STATE SCATTERING THEORY AND EIGENFUNCTION EXPANSIONS FOR A

In this lecture two families of generalized eigenfunctions for A are constructed and their completeness is proved. Physically, the generalized eigenfunctions are the steady-state acoustic waves which are produced when a plane wave is scattered by the obstacle $\Gamma = \mathbb{R}^n - \Omega$. Their construction is based on the limiting absorption theorem of Lecture 4. The eigenfunction expansions define two spectral representations for A. These provide explicit constructions of solutions in $L_2(\Omega)$ and solutions WFE of the d'Alembert equation which are the starting point for the asymptotic analysis of Lectures 7 and 8.

A generalized eigenfunction expansion for the operator A_0 is provided by the Plancherel theory of the Fourier transform. The functions

$$(6.1) \qquad w_0(x,p) = \frac{1}{(2\pi)^{n/2}} e^{ix\cdot p}, \quad x \in \mathbb{R}^n, \quad p \in \mathbb{R}^n$$

satisfy

$$(6.2) \qquad (\Delta + |p|^2) w_0(x,p) = 0 \quad \text{for all} \quad x \in \mathbb{R}^n, \; p \in \mathbb{R}^n.$$

Thus, formally, $A_0 w_0(x,p) = |p|^2 w_0(x,p)$ and $w_0(x,p)$ is a generalized eigenfunction of A_0 since $w_0(\cdot,p) \notin L_2(\mathbb{R}^n)$. The form of the eigenfunction expansion for

A_0 is given by the Plancherel formulas (2.19), (2.20) which may be written

(6.3) $$\hat{f}(p) = L_2(\mathbb{R}^n) - \lim_{M \to \infty} \int_{|x| \le M} \overline{w_0(x,p)}\, f(x)\,dx,$$

(6.4) $$f(x) = L_2(\mathbb{R}^n) - \lim_{M \to \infty} \int_{|p| \le M} w_0(x,p)\, \hat{f}(p)\,dp.$$

Equation (2.23) for an operator $\Psi(A_0)$ may be written

(6.5) $$\Psi(A_0)f(x) = L_2(\mathbb{R}^n) - \lim_{M \to \infty} \int_{|p| \le M} w_0(x,p)\Psi(|p|^2)\hat{f}(p)\,dp$$

which shows that (6.3) defines a spectral representation for A_0. This representation is generalized below to the operator A in an arbitrary exterior domain having the finite tiling property.

The function $w_0(p,x)$ is a steady-state solution of the d'Alembert equation which represents a plane wave propagating in the direction p if a time dependence $e^{-i|p|t}$ is assumed as in (4.3). The corresponding eigenfunction for A is defined to be the steady-state wave $w(x,p)$ which is produced when the obstacle Γ is immersed in the plane wave $w_0(x,p)$. Mathematically, this means that $w(x,p)$ should satisfy

(6.6) $$(\Delta + |p|^2)w(x,p) = 0 \quad \text{for} \quad x \in \Omega,$$

(6.7) $$D_\nu w(x,p) = 0 \qquad \text{for} \quad x \in \partial\Omega.$$

Moreover, if $w'(x,p)$ is defined by

(6.8) $$w(x,p) = w_0(x,p) + w'(x,p) \quad \text{for} \quad |x| \ge r_0$$

then $w'(x,p)$ will be required to satisfy the outgoing Sommerfeld radiation condition. Physically, this corresponds to the requirement that the total wave $w(x,p)$ should consist of the "incident wave" $w_0(x,p)$ plus an outgoing "scattered wave" $w'(x,p)$. Eigenfunctions of this form will be called "distorted plane waves", following T. Ikebe [14]. A second family of distorted plane waves is obtained by requiring that $w'(x,p)$ satisfy the incoming radiation condition. It will be shown in Lecture 7 that both families are useful in the study of solutions of the d'Alembert equation.

The incoming and outgoing distorted plane waves will be denoted by $w_+(x,p)$ and $w_-(x,p)$, respectively. For domains Ω with irregular boundaries (6.7) must be replaced by the generalized Neumann condition. This suggests the following definition (cf. (4.25)-(4.27)).

DEFINITION. Let Ω be an exterior domain. Then the *outgoing* (resp., *incoming*) *distorted plane wave* $w_+(x,p)$ (resp., $w_-(x,p)$) for Ω and $p \in \mathbb{R}^n$ is the function characterized by the conditions

(6.9) $$w_\pm(\cdot,p) \in L_2^{N,loc}(\Delta,\overline{\Omega}),$$

(6.10) $$(\Delta + |p|^2)w_\pm(x,p) = 0 \quad \text{for} \quad x \in \Omega,$$

(6.11) $\left\{\begin{array}{l}\text{the function } w_+(x,p) - w_0(x,p) \text{ (resp.,} \\ w_-(x,p) - w_0(x,p)) \text{ satisfies the outgoing} \\ \text{(resp., incoming) radiation condition.}\end{array}\right.$

Note that the functions $w_+(x,p)$ and $w_-(x,p)$ are unique if they exist, because the difference of two such functions would satisfy (6.9), (6.10) and the incoming or outgoing radiation condition and hence would be zero in Ω by Theorem 4.1. It will be shown next that the existence of the distorted plane waves follows from the limiting absorption theorem when Ω has the finite tiling property.

The construction of $w_\pm(x,p)$ will make use of the cut-off function

$$(6.12)\qquad j(x) = \phi_1(|x| - r_0), \quad x \in \mathbb{R}^n$$

where $\phi_1 \in C^\infty(\mathbb{R})$ is the function introduced in Lecture 4 and r_0 is chosen so that

$$(6.13)\qquad \Gamma = \mathbb{R}^n - \Omega \subset B(r_0).$$

Clearly $j \in C^\infty(\mathbb{R}^n)$, $0 \le j(x) \le 1$ and

$$(6.14)\qquad j(x) = \begin{cases} 0 & \text{for } |x| \le r_0 \\ 1 & \text{for } |x| \ge r_0 + 1. \end{cases}$$

In particular, $j(x)$ vanishes in a neighborhood of Γ. The eigenfunctions will be shown to have the form

$$(6.15)\qquad w_\pm(x,p) = j(x) w_0(x,p) + w_\pm'(x,p).$$

To find a way of constructing $w_\pm'(x,p)$ note that (6.10) and (6.15) imply that

$$(6.16)\qquad (\Delta + |p|^2) w_\pm'(x,p) = -(\Delta + |p|^2) j(x) w_0(x,p).$$

The right-hand side of (6.16) will be denoted by

$$(6.17)\quad M(x,p) = -(\Delta + |p|^2) j(x) w_0(x,p), \quad x \in \mathbb{R}^n, \quad p \in \mathbb{R}^n.$$

Note that

$$(6.18)\qquad M(x,p) \in C^\infty(\mathbb{R}^n \times \mathbb{R}^n)$$

and

$$\operatorname{supp} M(\cdot,p) \subset \{x: r_0 \le |x| \le r_0 + 1\} \quad \text{for all} \quad p \in \mathbb{R}^n.$$

In particular,

$$(6.19)\quad M(\cdot,p) \in L_2(\Omega_{r_0+1}) \subset L_2^{\text{vox}}(\bar{\Omega}) \quad \text{for all} \quad p \in \mathbb{R}^n.$$

This suggests that $w_\pm'(x,p)$ be defined by

$$(6.20)\quad w'(\cdot,p,z) = -R(z)M(\cdot,p), \quad p \in \mathbb{R}^n, \quad z \in \mathbb{C}_{1/2}^+$$

and

$$(6.21)\quad \begin{cases} w_\pm'(\cdot,p) = w'(\cdot,p,|p|^2 \pm i0) \\ \qquad\qquad = -R(|p|^2 \pm i0)M(\cdot,p), \quad p \in \mathbb{R}^n, \end{cases}$$

where $R(z) = (A - z)^{-1}$ for $z \in \mathbb{C}_{1/2}^+$ and the limits (6.21) exist in $L_2^{N,\text{loc}}(\Delta,\bar{\Omega})$ by the limiting absorption theorem. The notation

$$(6.22)\quad w(x,p,z) = j(x)w_0(x,p) + w'(x,p,z), \quad z \in \mathbb{C}_{1/2}^+,$$

will also be used, so that

$$(6.23)\quad w_\pm(\cdot,p) = w(\cdot,p,|p|^2 \pm i0) \in L_2^{N,\text{loc}}(\Delta,\bar{\Omega}).$$

The results of this construction are summarized as

THEOREM 6.1. *Let* Ω *be an exterior domain such that* $\Omega \in LC$. *Then for each* $p \in \mathbb{R}^n$ *there exists a unique outgoing distorted plane wave* $w_+(x,p)$ *and a unique incoming distorted plane wave* $w_-(x,p)$.

PROOF. The uniqueness of $w_\pm(x,p)$ follows from Theorem 4.1, as remarked above. The existence proof is based on the construction defined by (6.15), (6.20) and (6.21). It must be verified that $w_\pm$ satisfy (6.9), (6.10) and (6.11). Note that $j(\cdot)w_0(\cdot,p) \in L_2^{N,loc}(\Delta,\overline{\Omega})$, by direct calculation, and $w_\pm'(\cdot,p) \in L_2^{N,loc}(\Delta,\overline{\Omega})$ by Theorem 4.4. Thus $w_\pm$ satisfies (6.9). Also, Theorem 4.4 implies that

$$(6.24) \qquad (\Delta + |p|^2)w_\pm'(\cdot,p) = M(\cdot,p) \quad \text{in } \Omega.$$

Combining this with (6.17) and (6.15) gives (6.10). Finally, $w_\pm(x,p) - w_0(x,p) = w_\pm'(x,p)$ for $|x| \geq r_0 + 1$ and $w_\pm'(\cdot,p)$ satisfies the outgoing (+) or incoming (-) radiation condition by Theorem 4.4. This completes the proof of Theorem 6.1.

Note that w_+ and w_- are unique, as stated in Theorem 6.1, even though w_+' and w_-' are not unique since they depend on the choice of the cut-off function $j(x)$. The behavior of $w_\pm(x,p)$ at large distances from the boundary is described by

COROLLARY 6.2. *Under the hypotheses of Theorem* 6.1, *there exist functions* $\theta_\pm(\eta,p) \in C^\infty(S^{n-1} \times \mathbb{R}^n - \{0\})$ *such that*

$$(6.25)\quad \begin{cases} w_\pm(x,p) = w_0(x,p) + \dfrac{e^{\pm i|p||x|}}{|x|^{\frac{n-1}{2}}}\,\theta_\pm\left(\dfrac{x}{|x|},p\right) \\ \qquad\qquad + q_\pm(x,p) \end{cases}$$

where

$$(6.26)\qquad q_\pm(x,p) = O\left(\frac{1}{|x|^{\frac{n+1}{2}}}\right), \quad |x| \to \infty$$

uniformly for $\eta = x/|x| \in S^{n-1}$ *and* p *in any compact subset of* $\mathbb{R}^n - \{0\}$.

PROOF. Since $w_\pm(x,p) - w_0(x,p) = w'_\pm(x,p)$ for $|x| \geq r_0 + 1$ it is enough to consider $w'_\pm$. Now, application of Green's theorem and the boundary condition as in the proof of Lemma 4.6, equation (4.86), gives a representation

$$(6.27)\quad \begin{cases} w'_\pm(x,p) = \displaystyle\int_{S_r} \left\{ w'_\pm(x',p)\, \frac{\partial G_0(x,x',|p|^2 \pm i0)}{\partial |x'|} \right. \\ \qquad\qquad \left. - \dfrac{\partial w'_\pm(x',p)}{\partial |x'|}\, G_0(x,x',|p|^2 \pm i0) \right\} dS' \end{cases}$$

valid for $|x| > r > r_0 + 1$. Application of the classical asymptotic expansions for the Hankel functions [27] to G_0 gives the result (6.25), (6.26). This procedure gives an explicit formula for $\theta_\pm(\eta,p)$ as an integral over S_r which implies that $\theta_\pm \in C^\infty(S^{n-1} \times \mathbb{R}^n - \{0\})$. The formula will not be recorded here.

Physically, Corollary 6.2 states that $w_+(x,p)$ behaves for large $|x|$ like the sum of a plane wave $w_0(x,p)$ (the incident wave) and a diverging (+) or converging (−) spherical wave (the wave scattered by Γ). In acoustics $\theta_\pm(\eta,p)$ is called the *far-field amplitude*.

It will be shown next that each of the families $\{w_+(\cdot,p) : p \in \mathbb{R}^n\}$ and $\{w_-(\cdot,p) : p \in \mathbb{R}^n\}$ is a complete set of generalized eigenfunctions for A. The eigenfunction expansions will be derived from the spectral family $\{\Pi(\lambda)\}$ for A. The latter will be constructed by means of Stone's theorem, quoted above as Theorem 5.4. Since A is spectrally continuous equation (5.10) takes the form

$$(6.28)\qquad (f,\Pi(I)g) = \lim_{\sigma\to 0+} \frac{1}{2\pi i}\int_I (f,[R(\lambda+i\sigma) - R(\lambda-i\sigma)]g)\,d\lambda,$$

where $I = (a,b)$ is any interval and $f,g \in L_2(\Omega)$. To calculate the right-hand side of (6.28) note that the resolvent identity, that is, $R(z_1) - R(z_2) = (z_1 - z_2)R(z_1)R(z_2)$, implies that if $\operatorname{Im} z \neq 0$ then

$$(6.29)\qquad \left\{\begin{aligned} (f,[R(z)-R(\overline{z})]g) &= 2i \operatorname{Im} z(R(z)f,R(z)g) \\ &= 2i \operatorname{Im} z \int_\Omega \overline{R(z)f(x)}\,R(z)g(x)\,dx. \end{aligned}\right.$$

The spectral measure (6.28) will be calculated first for functions $f,g \in L_2^{\mathrm{vox}}(\overline{\Omega})$. Note that for such functions, if $j(x)$ is the cut-off function defined above, then

$$(6.30)\qquad \left\{\begin{aligned} &|(1 - j^2(x))\overline{R(z)f(x)}\,R(z)g(x)| \\ &\leq |R(z)f(x)|\,|R(z)g(x)|\,\chi_{r_0+1}(x), \end{aligned}\right.$$

where χ_{r_0+1} is the characteristic function of Ω_{r_0+1}, because $0 \le j(x) \le 1$ and $j(x) \equiv 1$ for $|x| \ge r_0 + 1$. Now $R(z)f$ and $R(z)g$ converge in $L_2(\Omega_{r_0+1})$ by the limiting absorption theorem. Thus

$$(6.31) \quad \begin{cases} \operatorname{Im} z \int_\Omega (1 - j^2(x)) \overline{R(z)f(x)} R(z)g(x)dx = \mathcal{O}(1), \\ \operatorname{Im} z \to 0, \end{cases}$$

where $\mathcal{O}(1)$ is a function of z which tends to zero when $\operatorname{Im} z \to 0$. Moreover, the convergence is uniform for $\operatorname{Re} z \in I$, by Lemma 4.7. Combining (6.29) and (6.31) gives

$$(6.32) \quad \begin{cases} (f, [R(z) - R(\bar{z})]g) \\ = 2i \operatorname{Im} z \displaystyle\int_\Omega \overline{j(x)R(z)f(x)}\, j(x)R(z)g(x)dx + \mathcal{O}(1). \end{cases}$$

Let

$$(6.33) \qquad J: L_2(\Omega) \to L_2(\mathbb{R}^n)$$

be the linear operator defined by

$$(6.34) \qquad (Jf)(x) = \begin{cases} j(x)f(x), & x \in \Omega \\ 0, & x \in \mathbb{R}^n - \Omega, \end{cases}$$

and note that J is bounded. In fact, taking $\operatorname{supp} f \subset \{x: j(x) = 1\}$ shows that $\|J\| = 1$. Then (6.32) implies

$$(6.35)\quad \begin{cases} (f,[R(z) - R(\overline{z})]g) \\ = 2i \text{ Im } z(JR(z)f,\ JR(z)g)_{L_2(\mathbb{R}^n)} + o(1) \\ = 2i \text{ Im } z((JR(z)f)^\wedge,\ (JR(z)g)^\wedge)_{L_2(\mathbb{R}^n)} + o(1). \end{cases}$$

In the last equation $(JR(z)f)^\wedge$ denotes the Fourier transform of $JR(z)f$ in $L_2(\mathbb{R}^n)$ and the equation follows from Parseval's formula. Equation (6.35) will be substituted in (6.28) and the limit evaluted. To this end define

$$(6.36)\quad \hat{f}(p,z) = \int_\Omega \overline{w(x,p,\overline{z})}f(x)dx,\ f \in L_2^{vox}(\overline{\Omega}),\ \text{Im } z \neq 0,$$

where $w(x,p,z)$ is defined by (6.20), (6.22). Note that for $p \in \mathbb{R}^n$ and $z \in \mathbb{C}_{\frac{1}{2}}^{+}$ fixed the integral in (6.36) converges pointwise. The connection with (6.35) is given by

LEMMA 6.3. *For all* $f \in L_2^{vox}(\overline{\Omega})$ *and* $z \in \mathbb{C}_+^{\frac{1}{2}}$, $\hat{f}(p,z)/(|p|^2 - z) \in L_2(\mathbb{R}^n)$ *and*

$$(6.37)\quad \hat{f}(p,z) = (|p|^2 - z)(JR(z)f)^\wedge(p).$$

PROOF. A heuristic proof may be based on the observation that, formally,

$$(6.38)\quad \begin{cases} (A-z)w(x,p,z) = (A-z)j(x)w_0(x,p) + (A-z)w'(x,p,z) \\ \qquad = M(x,p) + (|p|^2-z)j(x)w_0(x,p) - M(x,p) \\ \qquad = (|p|^2-z)j(x)w_0(x,p), \end{cases}$$

and hence

$$(6.39)\quad \begin{cases} \hat{f}(p,z) = \displaystyle\int_{\Omega} \overline{R(\bar{z})(A-\bar{z})w(x,p,\bar{z})}\, f(x)\,dx \\ \quad = \displaystyle\int_{\Omega} \overline{(|p|^2-\bar{z})\,j(x)\,w_0(x,p)\,R(z)}\, f(x)\,dx \\ \quad = (|p|^2-z)\displaystyle\int_{\Omega} \overline{w_0(x,p)}\, j(x)\,R(z)\,f(x)\,dx \\ \quad = (|p|^2-z)\,(JR(z)f)^{\wedge}(p)\,. \end{cases}$$

This argument is not rigorous because $w(\cdot,p,z)$ is not in $L_2(\Omega)$. A rigorous proof may be given as follows. (6.20), (6.22) and (6.36) imply that if $f \in L_2(\Omega_r) \subset L_2^{vox}(\bar{\Omega})$ then, for each $p \in \mathbb{R}^n$ and $z \in \mathbb{C}^+_{1/2}$,

$$(6.40)\quad \begin{cases} \hat{f}(p,z) = \displaystyle\int_{\Omega_r} \overline{w_0(x,p)}\, j(x)\,f(x)\,dx + \int_{\Omega_r} \overline{w'(x,p,\bar{z})}\, f(x)\,dx \\ = (Jf)^{\wedge}(p) - \displaystyle\int_{\Omega_r} \overline{R(\bar{z})M(\cdot,p)}\, f(x)\,dx \\ = (Jf)^{\wedge}(p) - \displaystyle\int_{\Omega_{r_0+1}} \overline{M(x,p)}\, R(z)\,f(x)\,dx \\ = (Jf)^{\wedge}(p) + \displaystyle\int_{\Omega_{r_0+1}} \overline{(\Delta+|p|^2)\{j(x)\,w_0(x,p)\}}\, R(z)\,f(x)\,dx. \end{cases}$$

The last step follows from the definition (6.17) of $M(x,p)$. The next-to-last step is valid because $M(\cdot,p) \in L_2(\Omega_{r_0+1}) \subset L_2(\Omega)$ and $f \in L_2(\Omega_r) \subset L_2(\Omega)$. To derive (6.37) from (6.40) it is necessary to integrate by parts in the last integral. To this end introduce the localizing function

$$(6.41) \qquad \phi_m(x) = \phi_1(m - |x|), \quad x \in \mathbb{R}^n,$$

which satisfies $\phi_m(x) \equiv 0$ for $|x| \geq m$ and $\phi_m(x) \equiv 1$ for $|x| \leq m - 1$. If $m - 1 \geq r_0 + 1$ then $\phi_m(x) \equiv 1$ on Ω_{r_0+1} and hence

$$(6.42) \quad \begin{cases} \hat{f}(p,z) = (Jf)^{\wedge}(p) \\ + \int_{\Omega} \overline{(\Delta+|p|^2)\{j(x)w_0(x,p)\}}\phi_m(x)R(z)f(x)dx. \end{cases}$$

Now $R(z)f \in L_2^N(\Delta,\Omega)$ which implies that $\phi_m R(z)f \in L_2^{1,\text{vox}}(\bar{\Omega})$. Moreover, $j(x)w_0(x,p) \in L_2^{N,\text{loc}}(\Delta,\bar{\Omega})$ because it is a smooth function which vanishes in a neighborhood of $\partial\Omega$. Thus the generalized Neumann condition (4.18) implies

$$(6.43) \quad \begin{cases} \int_{\Omega} \Delta\{j(x)\overline{w_0(x,p)}\}\phi_m(x)R(z)f(x)dx \\ = -\int_{\Omega} \nabla\{j(x)\overline{w_0(x,p)}\} \cdot \nabla\{\phi_m(x)R(z)f(x)\}dx. \end{cases}$$

Note that $j(x)w_0(x,p) \in L_2^1(\Omega_{m+1})$ and $\phi_m R(z)f \in L_2^N(\Delta,\Omega_{m+1})$ because $\phi_m(x) \equiv 0$ near $\{x: |x| = m+1\}$. Thus a second application of (4.18) gives

$$(6.44) \quad \begin{cases} \int_{\Omega} \Delta\{j(x)\overline{w_0(x,p)}\}\phi_m(x)R(z)f(x)dx \\ = \int_{\Omega} j(x)\overline{w_0(x,p)}\Delta\{\phi_m(x)R(z)f(x)\}dx. \end{cases}$$

Also, the rules for differentiating in $\mathcal{D}'(\Omega)$ imply that

$$(6.45)\quad \begin{cases} \Delta\{\phi_m(x)R(z)f(x)\} = \Delta\phi_m(x) \cdot R(z)f(x) \\ + 2\nabla\phi_m(x) \cdot \nabla R(z)f(x) + \phi_m(x)\Delta R(z)f(x). \end{cases}$$

Combining (6.42), (6.44), (6.45) and the equation $\Delta R(z)f = -f - zR(z)f$ gives

$$(6.46)\quad \begin{cases} \hat{f}(p,z) = (Jf)^{\wedge}(p) \\ + \int_{\Omega} j(x)\overline{w_0(x,p)}(\Delta+|p|^2)\{\phi_m(x)R(z)f(x)\}dx \\ = (Jf)^{\wedge}(p) + \int_{\Omega} j(x)\overline{w_0(x,p)}\{\Delta\phi_m \cdot R(z)f \\ + 2\nabla\phi_m \cdot \nabla R(z)f - \phi_m f - (z-|p|^2)\phi_m(x)R(z)f\}dx \\ = (Jf)^{\wedge}(p) - \int_{\Omega} \overline{w_0(x,p)}j(x)\phi_m(x)f(x)dx \\ + (|p|^2 - z)\int_{\Omega} \overline{w_0(x,p)}j(x)\phi_m(x)R(z)f(x)dx \\ + \int_{\Omega} \overline{w_0(x,p)}j(x)\{\Delta\phi_m(x) \cdot R(z)f(x) \\ + 2\nabla\phi_m(x) \cdot \nabla R(z)f(x)\}dx. \end{cases}$$

Note that $\phi_m(x) \equiv 1$ on supp f when $m - 1 \geq r$. Thus the first two terms on the right-hand side of (6.46) cancel and hence

$$(6.47)\quad \begin{cases} \dfrac{\hat{f}(p,z)}{(|p|^2-z)} = \displaystyle\int_\Omega \overline{w_0(x,p)}\,\phi_m(x)\,j(x)\,R(z)f(x)\,dx \\[2ex] + \dfrac{1}{(|p|^2-z)} \displaystyle\int_\Omega \overline{w_0(x,p)}\{\Delta\phi_m \cdot j(x)R(z)f(x) \\[2ex] + 2\nabla\phi_m \cdot j(x)\nabla R(x)f(x)\}dx = \Phi_n(\phi_m JR(z)f) \\[2ex] + \dfrac{1}{|p|^2-z}\,\Phi_n(\Delta\phi_m JR(z)f + 2\nabla\phi_m \cdot J\nabla R(z)f), \end{cases}$$

where Φ_n denotes the Fourier transform in $L_2(\mathbb{R}^n)$. It is clear from (6.47) that $\hat{f}(p,z)/(|p|^2-z) \in L_2(\mathbb{R}^n)$ since $(|p|^2-z)^{-1}$ is bounded and measurable on $\mathbb{R}^n$ when $\operatorname{Im} z \neq 0$. Moreover, $JR(z)f \in L_2(\mathbb{R}^n)$ and $J\nabla R(z)f \in L_2(\mathbb{R}^n)$, because $R(z)f \in L_2^N(\Delta,\Omega)$, and $\operatorname{supp}\Delta\phi_m \cup \operatorname{supp}\nabla\phi_m \subset \{x: m-1 \leq |x| \leq m\}$. It follows by Lebesgue's dominated convergence theorem that $\phi_m JR(z)f \to JR(z)f$ and $\Delta\phi_m JR(z)f + 2\nabla\phi_m \cdot J\nabla R(z)f \to 0$ in $L_2(\mathbb{R}^n)$ when $m \to \infty$. Finally, making $m \to \infty$ in (6.47) gives (6.37) because Φ_n is continuous on $L_2(\mathbb{R}^n)$.

Combining (6.35) and (6.37) gives

$$(6.48)\quad \begin{cases} (f,[R(z)-R(\overline{z})]g) \\[2ex] = 2i \operatorname{Im} z((|p|^2-z)^{-1}\hat{f}(p,z),(|p|^2-z)^{-1}\hat{g}(p,z))_{L_2(\mathbb{R}^n)} \\[2ex] + O(1) = 2i \operatorname{Im} z \displaystyle\int_{\mathbb{R}^n} \frac{\overline{\hat{f}(p,z)}\hat{g}(p,z)}{||p|^2-z|^2}\,dp + O(1). \end{cases}$$

Moreover, the term $\mathcal{O}(1)$ tends to zero uniformly for $\lambda = \mathrm{Re}\, z \in I$ when $\sigma = \mathrm{Im}\, z \to 0$ (Lemma 4.7). Hence, integration of (6.48) gives

$$(6.49)\quad \begin{cases} \dfrac{1}{2\pi i}\displaystyle\int_I (f,[R(\lambda+i\sigma) - R(\lambda-i\sigma)]g)\,d\lambda \\[2ex] = \displaystyle\int_I \int_{\mathbb{R}^n} \frac{\sigma}{\pi}\, \frac{\overline{\hat{f}(p,\lambda\pm i\sigma)}\,\hat{g}(p,\lambda\pm i\sigma)}{(\lambda-|p|^2)^2+\sigma^2}\, dp\,d\lambda + \mathcal{O}(1) \\[2ex] = \displaystyle\int_{\mathbb{R}^n} \left(\frac{\sigma}{\pi} \int_I \frac{\overline{\hat{f}(p,\lambda\pm i\sigma)}\,\hat{g}(p,\lambda\pm i\sigma)}{(\lambda-|p|^2)^2+\sigma^2}\, d\lambda \right) dp + \mathcal{O}(1)\,. \end{cases}$$

The second equation follows from the first by Fubini's theorem. Note that Lemma 6.3 implies that $\hat{f}(p,z)/(|p|^2-z)$ is a continuous $L_2(\mathbb{R}^n)$-valued function of $z \in \mathbb{C}^+_{\frac{1}{2}}$. Note also that (6.49) represents two equations corresponding to the choice of the $\pm$ sign. Both are obtained from (6.48) by the substitution $z = \lambda \pm i\sigma$.

Stone's theorem (6.28) and (6.49) will now be combined to obtain a representation of the spectral measure $(f,\Pi(I)g)$. The limit as $\sigma \to 0+$ of the inner integral in (6.49) can be evaluated by means of a well-known property of the Poisson kernel. The following formulation is due to E. C. Titchmarsh [40].

LEMMA 6.4. *Let* $\phi(\lambda)$, $\lambda \in \mathbb{R}$ *be a complex-valued function such that* $\phi(\lambda)/1+\lambda^2 \in L_1(\mathbb{R})$ *and assume that the limits* $\phi(a+)$ *and* $\phi(a-)$ *exist. Then*

$$(6.50)\qquad \lim_{\sigma\to 0+} \frac{\sigma}{\pi} \int_{\mathbb{R}} \frac{\phi(\lambda)}{(\lambda-a)^2+\sigma^2}\, d\lambda = \tfrac{1}{2}[\phi(a+) + \phi(a-)].$$

In order to apply this result to (6.49) some information concerning the point-wise continuity of $\hat{f}(p,\lambda\pm i\sigma)$ is needed. Note that for each fixed $p \in \mathbb{R}^n$ the mapping $\lambda\pm i\sigma \to w(\cdot,p,\lambda\pm i\sigma) \in L_2^{loc}(\overline{\Omega})$ is continuous for $\lambda \in I$, $0 \le \sigma \le \sigma_0$ by Lemma 4.7. In particular, for each $f \in L_2^{vox}(\overline{\Omega})$ the limits

$$(6.51)\qquad \hat{f}(p,\lambda\pm i0) = \int_{\Omega} \overline{w(x,p,\lambda\mp i0)}\, f(x)\,dx$$

exist for each $p \in \mathbb{R}^n$, $\lambda \in \mathbb{R}$. The following result will be proved.

LEMMA 6.5. *For each* $f \in L_2^{vox}(\overline{\Omega})$,

$$(6.52)\qquad \begin{cases} \hat{f}(p,\lambda\pm i\sigma) \text{ is continuous for all } p \in \mathbb{R}^n, \\ \lambda \in \mathbb{R} \text{ and } \sigma \ge 0. \end{cases}$$

PROOF. In view of the definition (6.36) of $\hat{f}(p,z)$ it will be sufficient to show that the mapping $(p,\lambda\pm i\sigma) \to w(\cdot,p,\lambda\pm i\sigma) \in L_2^{loc}(\overline{\Omega})$ is uniformly continuous for $p \in K$, $\lambda \in I$ and $0 < \sigma \le \sigma_0$ where $K \subset \mathbb{R}^n$ is any compact set, $I = [a,b]$ and $\sigma_0 > 0$. Moreover, in the definition (6.22) of $w(x,p,z)$ the term $j(x)w_0(x,p)$ is independent of z and is continuous for $(x,p) \in \mathbb{R}^n \times \mathbb{R}^n$ which implies that $p \to j(\cdot)w_0(\cdot,p) \in L_2^{loc}(\overline{\Omega})$ is uniformly continuous. Now consider $w'(\cdot,p,z) = -R(z)M(\cdot,p)$. The continuity of $M(x,p)$ on

$\mathbb{R}^n \times \mathbb{R}^n$ implies that $p \to M(\cdot,p) \in L_2(\Omega_{r_0+1})$ is uniformly continuous for $p \in K$. Moreover, if m is any positive constant then Lemma 4.6 implies that there is an $N = N(I,\sigma_0, m) > 0$ such that

$$(6.53)\quad \begin{cases} \|w'(\cdot,p,\lambda\pm i\sigma) - w'(\cdot,p',\lambda\pm i\sigma)\|_{L_2(\Omega_m)} \\ \leq N\|M(\cdot,p) - M(\cdot,p')\|_{L_2(\Omega_{r_0+1})} \end{cases}$$

for all $p,p' \in \mathbb{R}^n$, $\lambda \in I$ and $0 < \sigma \leq \sigma_0$. Combining this and the triangle inequality gives the estimate

$$(6.54)\quad \begin{cases} \|w'(\cdot,p,\lambda\pm i\sigma) - w'(\cdot,p',\lambda'\pm i\sigma')\|_{L_2(\Omega_m)} \\ \leq \|w'(\cdot,p,\lambda\pm i\sigma) - w'(\cdot,p',\lambda\pm i\sigma)\|_{L_2(\Omega_m)} \\ + \|w'(\cdot,p',\lambda\pm i\sigma) - w'(\cdot,p',\lambda'\pm i\sigma')\|_{L_2(\Omega_m)} \\ \leq N\|M(\cdot\, p) - M(\cdot,p')\|_{L_2(\Omega_{r_0+1})} \\ + \|w'(\cdot,p',\lambda\pm i\sigma) - w'(\cdot,p',\lambda'\pm i\sigma')\|_{L_2(\Omega_m)} \end{cases}$$

for all p and p' in $\mathbb{R}^n$, λ and λ' in I and $0 < \sigma, \sigma' \leq \sigma_0$. Thus the uniform continuity of $(p,\lambda\pm i\sigma) \to w(\cdot,p,\lambda\pm i\sigma) \in L_2^{loc}(\bar{\Omega})$ for $p \in K$, $\lambda \in I$, $0 < \sigma \leq \sigma$ follows from the uniform continuity of $p \to M(\cdot,p) \in L_2(\Omega_{r_0+1})$ for $p \in K$ and Lemma 4.7.

Lemma 6.5 implies that the functions $\hat{f}_+(p)$ and $\hat{f}_-(p)$ defined by

$$(6.55)\qquad \hat{f}_{\pm}(p) = \hat{f}(p,|p|^2 \mp i0), \quad p \in \mathbb{R}^n$$

are continuous on $\mathbb{R}^n$. Comparison with (6.23) shows that

$$(6.56)\qquad \hat{f}_{\pm}(p) = \int_{\Omega} \overline{w_{\pm}(x,p)}\, f(x)\,dx, \quad f \in L_2^{vox}(\overline{\Omega}).$$

These functions will be used to calculate the limit of the right-hand side of (6.49) for $\sigma \to 0+$. The limit of the inner integral is given by

LEMMA 6.6. *For each* $f \in L_2^{vox}(\overline{\Omega})$

$$(6.57)\qquad \left\{ \begin{aligned} &\lim_{\sigma \to 0+} \frac{\sigma}{\pi} \int_I \frac{\overline{\hat{f}(p,\lambda \pm i\sigma)}\,\hat{g}(p,\lambda \pm i\sigma)}{(\lambda - |p|^2)^2 + \sigma^2}\, d\lambda \\ &= \chi_I(|p|^2)\,\overline{\hat{f}_{\mp}(p)}\,\hat{g}_{\mp}(p) \quad \textit{for all} \quad p \in \mathbb{R}^n, \end{aligned} \right.$$

where

$$(6.58)\qquad \chi_I(\lambda) = \begin{cases} 1, & a < \lambda < b, \\ \tfrac{1}{2}, & \lambda = a \;\; \textit{and} \;\; \lambda = b \\ 0, & \lambda < a \;\; \textit{and} \;\; \lambda > b. \end{cases}$$

PROOF. Fix $p \in \mathbb{R}^n$ and write, for brevity,

$$(6.59)\qquad F_{\pm}(\lambda,\sigma) = \chi_I(\lambda)\,\overline{\hat{f}(p,\lambda \pm i\sigma)}\,\hat{g}(p,\lambda \pm i\sigma),$$

so

$$(6.60)\qquad F_{\pm}(\lambda,0) = \chi_I(\lambda)\,\overline{\hat{f}(p,\lambda \pm i0)}\,\hat{g}(p,\lambda \pm i0),$$

and

$$(6.61)\qquad F_{\pm}(|p|^2,0) = \chi_I(|p|^2)\overline{\hat{f}_{\mp}(p)}\hat{g}_{\mp}(p).$$

Now Lemma 6.5 implies that for each $\varepsilon > 0$ there exists a $\sigma_0 = \sigma_0(\varepsilon) > 0$ such that

$$(6.62)\qquad \begin{cases} |F_{\pm}(\lambda,\sigma) - F_{\pm}(\lambda,0)| \le \varepsilon \\ \text{for all } \lambda \in I \text{ and } 0 \le \sigma \le \sigma_0. \end{cases}$$

Moreover,

$$(6.63)\qquad \begin{cases} \dfrac{\sigma}{\pi}\displaystyle\int_I \frac{F_{\pm}(\lambda,\sigma)d\lambda}{(\lambda-|p|^2)^2+\sigma^2} \\ = \dfrac{\sigma}{\pi}\displaystyle\int_I \frac{F_{\pm}(\lambda,\sigma)-F_{\pm}(\lambda,0)}{(\lambda-|p|^2)^2+\sigma^2}\,d\lambda + \frac{\sigma}{\pi}\int_{\mathbb{R}} \frac{F_{\pm}(\lambda,0)}{(\lambda-|p|^2)^2+\sigma^2}\,d\lambda \\ = I_1(\sigma) + I_2(\sigma) \end{cases}$$

in an obvious notation. Now (6.62) implies that

$$(6.64)\qquad \begin{cases} |I_1(\sigma)| \le \dfrac{\sigma}{\pi}\displaystyle\int_I \frac{|F_{\pm}(\lambda,\sigma)-F_{\pm}(\lambda,0)|}{(\lambda-|p|^2)^2+\sigma^2}\,d\lambda \\ \le \varepsilon\left(\dfrac{\sigma}{\pi}\displaystyle\int_I \frac{d\lambda}{(\lambda-|p|^2)^2+\sigma^2}\right) = \varepsilon \end{cases}$$

for all $\sigma \le \sigma_0(\varepsilon)$; i.e., $\lim_{\sigma\to 0} I_1(\sigma) = 0$. Moreover, Lemma 6.4 implies that $\lim_{\sigma\to 0+} I_2(\sigma) = F_{\pm}(|p|^2,0)$. These results, with (6.63), imply (6.57).

Lemma 6.6, equation (6.49) and Stone's theorem suggest

THEOREM 6.7. *For every* f *and* g *in* $L_2^{vox}(\bar{\Omega})$ *and every bounded interval* $I \subset \mathbb{R}$

$$(6.65) \qquad (f,\Pi(I)g) = \int_{\mathbb{R}^n} \chi_I(|p|^2)\overline{\hat{f}_\pm(p)}\,\hat{g}_\pm(p)\,dp.$$

PROOF. Note that the integral on the right-hand side of (6.65) is finite because $\hat{f}_\pm$ and $\hat{g}_\pm$ are continuous and $\chi_I(|p|^2)$ has compact support. Moreover, (6.65) follows from (6.49) and (6.57) if passage to the limit under the integral sign in (6.49) is valid. The correctness of this interchange of limits will be deduced from the Lebesgue dominated convergence theorem and the following estimate.

LEMMA 6.8. *For each* $f \in L_2^{vox}(\bar{\Omega})$ *the function* $\hat{f}(\cdot,\lambda\pm i\sigma) \in L_2(\mathbb{R}^n)$ *for every* $\lambda \in \mathbb{R}$ *and* $\sigma \geq 0$. *Moreover, for each bounded interval* $I \subset \mathbb{R}$ *and* $\sigma_0 > 0$ *there exists a constant* $C = C(f,I,\sigma_0)$ *such that*

$$(6.66) \qquad \left\{ \begin{array}{l} \displaystyle\int_{\mathbb{R}^n} |\hat{f}(p,\lambda\pm i\sigma)|^2 dp \leq C \\ \text{for all } \lambda \in I \text{ and } 0 \leq \sigma \leq \sigma_0. \end{array} \right.$$

PROOF. The starting point is equation (6.40) above which can be written

$$(6.67) \qquad \hat{f}(p,z) = (Jf)^\wedge(p) + g(p,z), \quad p \in \mathbb{R}^n, \quad z \in \mathbb{C}^+_{\frac{1}{2}},$$

where

$$(6.68)\qquad g(p,z) = -\int_{\Omega_{r_0+1}} \overline{M(x,p)}\, R(z) f(x)\, dx.$$

This implies that

$$(6.69)\qquad |\hat{f}(p,z)|^2 \leq 2(|(Jf)^\wedge(p)|^2 + |g(p,z)|^2).$$

Moreover,

$$(6.70)\qquad \int_{\mathbb{R}^n} |(Jf)^\wedge(p)|^2 dp = \int_{\mathbb{R}^n} |Jf(x)|^2 dx \leq \|f\|^2 .$$

Hence it will suffice to prove (6.66) with $\hat{f}$ replaced by g. Now

$$(6.71)\qquad -M(x,p) = (\Delta j(x) + 2ip \cdot \nabla j(x))\, w_0(x,p).$$

Thus

$$(6.72)\qquad g(p,z) = \sum_{k=0}^{n} g_k(p,z),$$

where

$$(6.73)\qquad g_0(p,z) = \int_\Omega \overline{w_0(x,p)}\, \Delta j(x) R(z) f(x)\, dx,$$

and

$$(6.74)\qquad \begin{cases} g_k(p,z) = -2ip_k \displaystyle\int_\Omega \overline{w_0(x,p)}\, D_k j(x) R(z) f(x)\, dx, \\ k = 1, 2, \ldots, n. \end{cases}$$

Parseval's formula and Lemma 4.6 imply that

$$(6.75)\quad \begin{cases} \displaystyle\int_{\mathbb{R}^n} |g_0(p,\lambda\pm i\sigma)|^2 dp = \|(\Delta j)R(\lambda\pm i\sigma)f\|^2_{L_2(\Omega_{r_0+1})} \\ \le C^2 \|R(\lambda\pm i\sigma)f\|^2_{L_2(\Omega_{r_0+1})} \le C^2M^2 \|f\|^2_{L_2(\Omega_r)} \end{cases}$$

for all $\lambda \in I$, $0 \le \sigma \le \sigma_0$ where C is a bound for $|\Delta j(x)|$ and $M = M(I,\sigma_0, r_0+1,r)$ is the constant of Lemma 4.6. To find a similar estimate for $g_k(p,z)$ with $k = 1, 2, \ldots, n$ note that

$$(6.76)\quad ip_k w_0(x,p)D_k j(x) = D_k\{w_0(x,p)D_k j(x)\} - w_0(x,p)D_k^2 j(x).$$

Substituting this in (6.74) gives

$$(6.77)\quad \begin{cases} g_k(p,z) = 2\displaystyle\int_\Omega D_k\{\overline{w_0(x,p)}D_k j(x)\}R(z)f(x)\,dx \\ \qquad - 2\displaystyle\int_\Omega \overline{w_0(x,p)}D_k^2 j(x)R(z)f(x)\,dx \\ \qquad = g_k^1(p,z) + g_k^2(p,z). \end{cases}$$

Note that for each fixed $p \in \mathbb{R}^n$ the function $\overline{w_0(x,p)}D_k j(x)$ is in $\mathcal{D}(\Omega)$. Moreover, $R(z)f \in L_2^N(\Delta,\Omega)$ and hence has derivatives $D_k R(z)f \in L_2(\Omega)$. Thus, the distributional definition of derivatives implies that

$$(6.78)\quad g_k^1(p,z) = -2\int_\Omega \overline{w_0(x,p)}D_k j(x)D_k R(z)f(x)\,dx.$$

Proceeding as in the case of $g_0(p,z)$ gives

$$(6.79)\quad \left\{\begin{aligned} \int_{\mathbb{R}^n} |g_k^1(p,\lambda\pm i\sigma)|^2 dp &= \|D_k j D_k R(\lambda\pm i\sigma) f\|^2_{L_2(\Omega_{r_0+1})} \\ &\le C_k^2 \|D_k R(\lambda\pm i\sigma) f\|^2_{L_2(\Omega_{r_0+1})} \\ &\le C_k^2 \|R(\lambda\pm i\sigma) f\|^2_{L_2^1(\Delta,\Omega_{r_0+1})} \\ &\le C_k^2 M^2 \|f\|^2_{L_2(\Omega_r)} \end{aligned}\right.$$

for all $\lambda \in I$, $0 \le \sigma \le \sigma_0$ where $C_k = \max|D_k j(x)|$. Finally, $g_k^2(p,z)$ has the same form as $g_0(p,z)$ and hence satisfies an estimate like (6.75). Combining these results gives the estimate (6.66).

PROOF OF THEOREM 6.7. Note that it is enough to prove (6.65) for the case where $g = f$. The general case then follows by polarization. Thus only the case where $g = f$ will be considered.

In equation (6.49) the limit of the left-hand side for $\sigma \to 0+$ is $(f, \Pi(I) f)$ when $g = f$, by Stone's theorem, and the term $\mathcal{O}(1)$ tends to zero with σ. The integral on the right has the form

$$(6.80)\quad \int_{\mathbb{R}^n} I(p,\sigma)\, dp = \int_{|p|\le m} I(p,\sigma)\, dp + \int_{|p|\ge m} I(p,\sigma)\, dp,$$

where

$$(6.81)\quad I(p,\sigma) = \frac{\sigma}{\pi} \int_I \frac{|\hat{f}(p,\lambda\pm i\sigma)|^2}{(\lambda-|p|^2)^2+\sigma^2}\, d\lambda, \quad p \in \mathbb{R}^n, \ \sigma > 0.$$

Choose m in (6.80) so large that $|\lambda - |p|^2| \geq 1$ for all $|p| \geq m$ and hence $\chi_I(|p|^2) = 0$ for $|p| \geq m$. Next note that the continuity of $\hat{f}(p,\lambda \pm i\sigma)$, proved in Lemma 6.5, implies that there is a constant C_0 such that

$$(6.82) \quad \begin{cases} |\hat{f}(p,\lambda \pm i\sigma)|^2 \leq C_0 \\ \text{for all } |p| \leq m, \ \lambda \in I, \ 0 < \sigma \leq \sigma_0. \end{cases}$$

Thus for $|p| \leq m$, $0 < \sigma \leq \sigma_0$,

$$(6.83) \qquad I(p,\sigma) \leq C_0 \left(\frac{\sigma}{\pi} \int_I \frac{d\lambda}{(\lambda - |p|^2)^2 + \sigma^2} \right) \leq C_0 .$$

It follows by Lebesgue's bounded convergence theorem and Lemma 6.6 that

$$(6.84) \quad \begin{cases} \lim\limits_{\sigma \to 0+} \displaystyle\int_{|p| \leq m} I(p,\sigma)\, dp = \int_{|p| \leq m} \chi_I(|p|^2) |\hat{f}_{\mp}(p)|^2 dp \\ \qquad\qquad = \displaystyle\int_{\mathbb{R}^n} \chi_I(|p|^2) |\hat{f}_{\mp}(p)|^2 dp , \end{cases}$$

because $\chi_I(|p|^2) \equiv 0$ for $|p| \geq m$. Next, note that

$$(6.85) \qquad 0 < \frac{\sigma}{\pi} \frac{1}{(\lambda - |p|^2)^2 + \sigma^2} \leq \frac{\sigma}{\pi} \quad \text{for all } |p| \geq m ,$$

and hence by Lemma 6.8

$$(6.86)\quad \begin{cases} \displaystyle\int_{|p|\geq m} I(p,\sigma)\,dp \\ \displaystyle = \int_I \int_{|p|\geq m} \left(\frac{\sigma}{\pi}\,\frac{1}{(\lambda-|p|^2)^2+\sigma^2}\right) |\hat{f}(p,\lambda\pm i\sigma)|^2 dp\,d\lambda \\ \displaystyle \leq \frac{\sigma}{\pi} \int_I \int_{|p|\geq m} |\hat{f}(p,\lambda\pm i\sigma)|^2 dp\,d\lambda \leq \frac{|I|C}{\pi}\,\sigma, \end{cases}$$

where C is the constant of Lemma 6.8 and $|I|$ is the length of I. It follows that

$$(6.87)\qquad \lim_{\sigma\to 0+} \int_{|p|\geq m} I(p,\sigma)\,dp = 0.$$

Combining (6.80), (6.84) and (6.87) gives

$$(6.88)\qquad \lim_{\sigma\to 0+} \int_{\mathbb{R}^n} I(p,\sigma)\,dp = \int_{\mathbb{R}^n} \chi_I(|p|^2)\,|\hat{f}_{\mp}(p)|^2 dp,$$

which completes the proof of Theorem 6.7.

Theorem 6.7 provides the key to the relationship between the spectral family $\{\Pi(\lambda)\}$ of A and the generalized eigenfunctions $w_+(x,p)$ and $w_-(x,p)$. It will now be used to develop a complete analogue of the Plancherel theory for exterior domains having the finite tiling property. The functions $\hat{f}_+(p)$ and $\hat{f}_-(p)$, given by (6.56) when $f \in L_2^{\text{vox}}(\overline{\Omega})$, play the role of the generalized Fourier transform. It will be shown next that they are always in $L_2(\mathbb{R}^n)$.

LEMMA 6.9. *For all* $f \in L_2^{vox}(\overline{\Omega})$ *the functions* $\hat{f}_{\pm}(p) \in L_2(\mathbb{R}^n)$ *and*

$$(6.89) \qquad \|\hat{f}_{\pm}\|_{L_2(\mathbb{R}^n)} = \|f\|_{L_2(\Omega)} .$$

PROOF. Apply Theorem 6.7 with $g = f$ and $I = (-1,\lambda)$. Then $\Pi(I) = \Pi(\lambda)$, since $\sigma(A) \subset \overline{\mathbb{R}_+}$, and hence since $\Pi(\lambda)^* = \Pi(\lambda) = \Pi(\lambda)^2$,

$$(6.90) \quad \|\Pi(\lambda) f\|^2_{L_2(\Omega)} = (f,\Pi(\lambda) f)_{L_2(\Omega)} = \int_{|p|\le\sqrt{\lambda}} |\hat{f}_{\pm}(p)|^2 dp$$

for all $\lambda \geq 0$. Now make $\lambda \to +\infty$. The left-hand side of (6.90) converges to $\|f\|^2$ by a basic property of spectral families. Hence the right-hand side has a finite limit, which proves that $\hat{f}_{\pm} \in L_2(\mathbb{R}^n)$, and the limiting form of (6.90) is (6.89).

The next problem is to extend the definition of $\hat{f}_{\pm}$ from $L_2^{vox}(\overline{\Omega})$ to all of $L_2(\Omega)$. This is done by

LEMMA 6.10. *For all* $f \in L_2(\Omega)$ *the limits*

$$(6.91) \qquad \hat{f}_{\pm}(p) = \underset{M \to \infty}{L_2(\mathbb{R}^n)\text{-lim}} \int_{\Omega_M} \overline{w_{\pm}(x,p)}\, f(x)\, dx$$

exist and, with this extended definition of $\hat{f}_{\pm}$, *equation* (6.89) *holds for all* $f \in L_2(\Omega)$.

PROOF. Note that (6.91) coincides with (6.56) when $f \in L_2^{vox}(\overline{\Omega})$ and hence provides an extension of that definition. To prove the existence of the limit (6.91) when $f \in L_2(\Omega)$ define

$$(6.92) \qquad f_M(x) = \begin{cases} f(x), & x \in \Omega_M = \Omega \cap \{x: |x| < M\} \\ 0, & x \in \Omega - \Omega_M, \end{cases}$$

and note that $\lim_{M\to\infty} f_M = f$ in $L_2(\Omega)$. Apply (6.89) to $f_M - f_N \in L_2^{vox}(\bar{\Omega})$:

$$(6.93) \qquad \|\hat{f}_{M\pm} - \hat{f}_{N\pm}\|_{L_2(\mathbb{R}^n)} = \|f_M - f_N\|_{L_2(\Omega)}$$

for all finite M and N. The right-hand side tends to zero when $M, N \to \infty$ in any way because $\{f_M\}$ is convergent in $L_2(\Omega)$. Hence $\{\hat{f}_{M\pm}\}$ is a Cauchy sequence in $L_2(\mathbb{R}^n)$ and hence converges to a limit $\hat{f}_\pm \in L_2(\mathbb{R}^n)$ by the completeness of $L_2(\mathbb{R}^n)$. This proves (6.91) because

$$(6.94) \qquad \hat{f}_{M\pm}(p) = \int_{\Omega_M} \overline{w_\pm(x,p)}\, f(x)\, dp.$$

Finally, (6.89) is extended to arbitrary $f \in L_2(\Omega)$ by taking $N = 0$ in (6.93); i.e. $f_N = 0$, $\hat{f}_{N\pm} = 0$, and making $M \to \infty$.

COROLLARY 6.11. *The representation* (6.65) *of the spectral measure* $\Pi(I)$ *holds for all* f *and* g *in* $L_2(\Omega)$ *and all* $I \subset \mathbb{R}$.

PROOF. Write (6.65) for f_M and g_M and make $M \to \infty$.

COROLLARY 6.12. *For all* $f \in L_2(\Omega)$ *and all* $\lambda \in \mathbb{R}$

$$(6.95) \qquad \Pi(\lambda) f(x) = \begin{cases} \int_{|p| \le \sqrt{\lambda}} w_\pm(x,p)\, \hat{f}_\pm(p)\, dp, & \lambda \ge 0, \\ 0, & \lambda < 0 \end{cases}$$

in $L_2(\Omega)$. *In particular, the right-hand side of* (6.95) *defines a function in* $L_2(\Omega)$ *for each* $\lambda \in \mathbb{R}$.

PROOF. Note first that $w_\pm(x,p) = j(x)w_0(x,p) + w_\pm'(x,p)$ where $(\Delta + |p|^2)w_\pm'(x,p) = M(x,p)$ for all $x \in \Omega$, $p \in \mathbb{R}^n$. Now $M(x,p) \in C^\infty(\mathbb{R}^n \times \mathbb{R}^n)$ and $\operatorname{supp} M(\cdot,p) \subset \{x: r_0 \le |x| \le r_0+1\}$ is disjoint from $\Gamma = \mathbb{R}^n - \Omega$. It follows easily from standard regularity theory for Δ that $w_\pm(x,p) \in C^\infty(\Omega \times \mathbb{R}^n)$. Also, $\hat{f}_\pm \in L_2(\mathbb{R}^n)$ and hence $\hat{f}_\pm \in L_1(K)$ where K is any compact set. Thus the right-hand side of (6.95) is finite for each $x \in \Omega$. To prove equation (6.95) note that (6.65) with $f \in L_2(\Omega)$, $g \in \mathcal{D}(\Omega)$ and $I = (-1,\lambda)$ implies

$$
(6.96)\quad \left\{
\begin{aligned}
(g,\Pi(\lambda)f) &= \int_{|p|\le\sqrt{\lambda}} \overline{\hat{g}_\pm(p)}\,\hat{f}_\pm(p)\,dp \\
&= \int_{|p|\le\sqrt{\lambda}} \left(\int_\Omega w_\pm(x,p)\,\overline{g(x)}\,dx\right)\hat{f}_\pm(p)\,dp \\
&= \int_\Omega \overline{g(x)} \left(\int_{|p|\le\sqrt{\lambda}} w_\pm(x,p)\,\hat{f}_\pm(p)\,dp\right)dx,
\end{aligned}
\right.
$$

where the last step follows by Fubini's theorem. (6.96) implies that the left- and right-hand sides of (6.95) coincide as distributions in $\mathcal{D}'(\Omega)$. In particular, the right-hand side of (6.95) coincides with a distribution in $L_2(\Omega)$.

Corollary 6.12 implies that every $f \in L_2(\Omega)$ has an eigenfunction expansion in $L_2(\Omega)$.

COROLLARY 6.13. *For all* $f \in L_2(\Omega)$

$$f(x) = \underset{M \to \infty}{L_2(\Omega)\text{-lim}} \int_{|p|\le M} w_\pm(x,p)\hat{f}_\pm(p)\,dp. \tag{6.97}$$

This follows immediately from (6.95) and a basic property of the spectral family. Note the (6.97) defines two representations of f, one with $w_+(x,p)$ and one with $w_-(x,p)$.

It was remarked in Lecture 2 that the Fourier transform defines an operator $\Phi: L_2(\mathbb{R}^n) \to L_2(\mathbb{R}^n)$ by $\Phi f = \hat{f}$ (the subscript n will be suppressed in what follows) and that Φ is unitary and provides a spectral representation for the operator A_0. These results will now be generalized to exterior domains $\Omega \in LC$. To this end define operators

$$\Phi_\pm : L_2(\Omega) \to L_2(\mathbb{R}^n), \tag{6.98}$$

by

$$\Phi_\pm f = \hat{f}_\pm \quad \text{for all} \quad f \in L_2(\Omega). \tag{6.99}$$

Then the following generalization of the Plancherel theory holds.

THEOREM 6.14. *The operators* Φ_+ *and* Φ_- *are unitary; that is,*

$$\|\Phi_\pm f\|_{L_2(\mathbb{R}^n)} = \|f\|_{L_2(\Omega)} \quad \textit{for all} \quad f \in L_2(\Omega) \tag{6.100}$$

and

$$\Phi_{\pm} L_2(\Omega) = L_2(\mathbb{R}^n) \tag{6.101}$$

or, equivalently,

$$\Phi_{\pm}^{*} \Phi_{\pm} = 1 \quad \textit{and} \quad \Phi_{\pm} \Phi_{\pm}^{*} = 1. \tag{6.102}$$

THEOREM 6.15. *Let* $\Psi(\lambda)$ *be a bounded, Lebesgue measurable function of* $\lambda \geq 0$. *Then* $\Psi(A)$, *defined by the spectral theorem, is a bounded operator on* $L_2(\Omega)$ *and*

$$\Phi_{\pm} \Psi(A) = \Psi(|\cdot|^2) \Phi_{\pm} \tag{6.103}$$

where $\Psi(|\cdot|^2)$ *denotes the operation of multiplying by* $\Psi(|p|^2)$ *in* $L_2(\mathbb{R}^n)$.

Theorem 6.15 will be proved first and then used in the proof of Theorem 6.14.

PROOF OF THEOREM 6.15. Note first that it will be enough to prove that

$$(\Phi_{\pm} Au)(p) = |p|^2 (\Phi_{\pm} u)(p) \quad \text{for all} \quad u \in D(A). \tag{6.104}$$

Equation (6.103) then follows by standard methods of approximation; see [18, pp. 530-531]. Next it is enough to verify (6.104) for all functions in a "core" of A; i.e., a set $D \subset D(A)$ such that $\{(u, Au) : u \in D\}$ is dense in the graph of A. It will be shown that

$$D = D(A) \cap L_2^{\text{VOX}}(\overline{\Omega}) \tag{6.105}$$

is a core of A. To this end let $u \varepsilon D(A)$ and define $u_m = \phi_m u$ where $\phi_m(x) = \phi_1(m - |x|)$ is the function (6.41) used in the proof of Lemma 6.3. It will be shown that $u_m \varepsilon D$ for $m > r_0 + 1$ and $u_m \to u$ in the graph norm of A. Recall that $\phi_m(x) \equiv 0$ for $|x| \geq m$, $\phi_m(x) \equiv 1$ for $|x| \leq m-1$, $0 \leq \phi_m(x) \leq 1$ and $\phi_m \varepsilon \mathcal{D}(\mathbb{R}^n)$. Thus

$$(6.106) \quad \begin{cases} \nabla u_m = \phi_m \nabla u + u \nabla \phi_m \\ \\ \Delta u_m = \phi_m \Delta u + 2\nabla\phi_m \cdot \nabla u + u\Delta\phi_m \end{cases}$$

which imply that $u_m \varepsilon L_2^1(\Delta,\Omega)$. Moreover, u_m satisfies the generalized Neumann condition (3.10) because u satisfies (3.10) and $\phi_m(x) \equiv 1$ near $\partial\Omega$. To see this note that if $w \varepsilon L_2^1(\Omega)$ then (6.106) implies, after some rearrangement of terms,

$$(6.107) \quad \begin{cases} \displaystyle\int_\Omega \{(\Delta u_m) w + \nabla u_m \cdot \nabla w\} dx \\ \displaystyle\qquad = \int_\Omega \{(\Delta u)\phi_m w + \nabla u \cdot \nabla(\phi_m w)\} dx \\ \displaystyle\qquad + \int_\Omega \{(uw)\Delta\phi_m + \nabla(uw) \cdot \nabla\phi_m\} dx. \end{cases}$$

The first term on the right in (6.107) vanishes because u satisfies (6.10) and $\phi_m w \varepsilon L_2^1(\Omega)$. The second term vanishes by the distribution theory definition of derivative:

$$(6.108) \qquad \int_\Omega \{(uw) D_j^2 \psi + D_j(uw) D_j \psi\} dx = 0$$

for all $\psi \varepsilon \mathcal{D}(\Omega)$. It is easy to verify that $\nabla(uw) =$

$u\nabla w + w\nabla u \in L_1(\Omega)$. Hence the last integral in the last integral in (6.107) vanishes because $\phi_m \in \mathcal{D}(\mathbb{R}^n)$ and supp $\nabla\phi_m$ is a compact subset of Ω. Finally, the fact that $u_m \to u$ and $Au_m = -\Delta u_m \to Au = -\Delta u$ in $L_2(\Omega)$ is obvious from (6.106) and the properties of ϕ_m.

Next, (6.104) will be verified for $u \in D$. To this end note that

$$(6.109)\qquad \begin{cases} (\Phi_\pm Au)(p) = \displaystyle\int_\Omega \overline{w_\pm(x,p)}\, Au(x)\, dx \\ \qquad\qquad\quad = \displaystyle\int_\Omega \overline{\phi_m(x)\, w_\pm(x,p)}\, Au(x)\, dx, \end{cases}$$

where m is chosen so that $\phi_m(x) \equiv 1$ on supp u. Now $w_\pm(\cdot,p) \in L_2^{N,loc}(\Delta,\overline{\Omega})$ and hence $\phi_m w_\pm(\cdot,p) \in L_2^N(\Delta,\Omega) = D(A)$ by the argument given above. Moreover,

$$(6.110)\qquad \begin{cases} A(\phi_m(x) w_\pm(x,p)) = -\Delta(\phi_m(x) w_\pm(x,p)) \\ = -\phi_m(x)\, \Delta w_\pm(x,p) - 2\nabla\phi_m(x) \cdot \nabla w_\pm(x,p) \\ \qquad - \Delta\phi_m(x)\, w_\pm(x,p) \\ = -\Delta w_\pm(x,p) \quad \text{for all } x \in \text{supp } u, \end{cases}$$

because $\phi_m(x) \equiv 1$ on supp u. Also $\Delta w_\pm + |p|^2 w_\pm = 0$. Thus (6.109) implies

$$(6.111)\qquad \begin{cases} (\Phi_\pm Au)(p) = (\phi_m w_\pm(\cdot,p), Au)_{L_2(\Omega)} \\ \qquad\qquad\quad = (A(\phi_m w_\pm(\cdot,p)), u)_{L_2(\Omega)} \\ \qquad\qquad\quad = |p|^2 \displaystyle\int_\Omega \overline{w_\pm(x,p)}\, u(x)\, dx = |p|^2 (\Phi_\pm u)(p) \end{cases}$$

for all $p \in \mathbb{R}^n$. Finally, replacing u by u_m in (6.111) and making $m \to \infty$ gives (6.104) because $u_m \to u$ and $Au_m \to Au$ in $L_2(\Omega)$ and $\Phi_\pm$ is continuous.

PROOF OF THEOREM 6.14. Equation (6.100) states that equation (6.89) holds for all $f \in L_2(\Omega)$ and was proved in Lemma 6.10. It is also equivalent to the first of equations (6.102). Equation (6.101) is equivalent to the second equation of (6.102). It does not follow from the spectral theorem for A. The proof of it given below is based on Theorem 6.15 and the following two lemmas.

LEMMA 6.16. $\Phi_\pm \Phi_\pm^* = 1$ *if and only if*

$$(6.112) \qquad N(\Phi_\pm^*) \equiv \{h : \Phi_\pm^* h = 0\} = \{0\}.$$

LEMMA 6.17. *For all* $h \in L_2(\mathbb{R}^n)$

$$(6.113) \qquad \Phi_\pm^* h(x) = L_2(\Omega) - \lim_{M \to \infty} \int_{|p| \le M} w_\pm(x,p) h(p)\, dp.$$

PROOF OF LEMMA 6.16. It is clear that $\Phi_\pm \Phi_\pm^* = 1$ implies (6.112). To prove the converse let $f \in L_2(\mathbb{R}^n)$ and define

$$(6.114) \qquad h = (\Phi_\pm \Phi_\pm^* - 1) f \in L_2(\mathbb{R}^n).$$

Applying $\Phi_\pm^*$ to (6.114) and using the first equation of (6.102) gives

$$(6.115) \qquad \Phi_\pm^* h = ((\Phi_\pm^* \Phi_\pm) \Phi_\pm^* - \Phi_\pm^*) f = (\Phi_\pm^* - \Phi_\pm^*) f = 0.$$

Thus (6.112) implies that $h = 0$; i.e. $(\Phi_+ \Phi_+^* - 1) f = 0$

for all $f \in L_2(\mathbb{R}^n)$ and hence $\Phi_\pm \Phi_\pm^* = 1$.

PROOF OF LEMMA 6.17. Let $h \in L_2(\mathbb{R}^n)$ and define

$$(6.116) \qquad h_M(p) = \begin{cases} h(p), & |p| \le M, \\ 0, & |p| > M. \end{cases}$$

Then if $f \in L_2^{\text{vox}}(\overline{\Omega})$ the definition of $\Phi_\pm^*$ implies that

$$(6.117) \qquad (f, \Phi_\pm^* h_M)_{L_2(\Omega)} = (\Phi_\pm f, h_M)_{L_2(\mathbb{R}^n)} .$$

Writing the last scalar product as an integral and substituting the representation (6.56) for $\Phi_\pm f = \hat{f}_\pm$ gives

$$(6.118) \qquad \begin{cases} (f, \Phi_\pm^* h_M)_{L_2(\Omega)} = \displaystyle\int_{|p|\le M} \left(\int_\Omega w_\pm(x,p) \overline{f(x)}\, dx \right) h(p)\, dp \\ \qquad = \displaystyle\int_\Omega \overline{f(x)} \left(\int_{|p|\le M} w_\pm(x,p) h(p)\, dp \right) dx \end{cases}$$

by Fubini's theorem. This implies that

$$(6.119) \qquad \Phi_\pm^* h_M(x) = \int_{|p|\le M} w_\pm(x,p) h(p)\, dp \in L_2(\Omega)$$

since $L_2^{\text{vox}}(\overline{\Omega})$ is dense in $L_2(\Omega)$. Note that, in particular, this proves that the right-hand side of (6.119) is in $L_2(\Omega)$. Finally, if $M \to \infty$ then $h_M \to h$ in $L_2(\mathbb{R}^n)$ and hence $\Phi_\pm^* h_M \to \Phi_\pm^* h$, since $\Phi_\pm^*$ is bounded. Thus (6.119) implies (6.113).

PROOF OF THEOREM 6.14 (concluded). By Lemma 6.16 it will be enough to verify (6.112). To this end let

$h \in N(\Phi_\pm^*)$; i.e., $h \in L_2(\mathbb{R}^n)$ and

$$\Phi_\pm^* h = 0. \tag{6.120}$$

It follows that if $\Psi(\lambda)$ is any bounded Lebesgue measurable function on $\lambda \geq 0$ then

$$h'(p) = \Psi(|p|^2) h(p) \in N(\Phi_\pm^*). \tag{6.121}$$

Indeed, if $f \in L_2(\Omega)$ then Theorem 6.15 implies that

$$\left\{\begin{aligned} (f, \Phi_\pm^* h')_{L_2(\Omega)} &= (\Phi_\pm f, h')_{L_2(\mathbb{R}^n)} \\ &= \int_{\mathbb{R}^n} \overline{\hat{f}_\pm(p)} \Psi(|p|^2) h(p)\, dp = \int_{\mathbb{R}^n} \overline{\overline{\Psi}(|p|^2) \hat{f}_\pm(p)} h(p)\, dp \\ &= (\overline{\Psi}(|\cdot|^2) \Phi_\pm f, h)_{L_2(\mathbb{R}^n)} = (\Phi_\pm \overline{\Psi}(A) f, h)_{L_2(\mathbb{R}^n)} \\ &= (\overline{\Psi}(A) f, \Phi_\pm^* h)_{L_2(\Omega)} = 0 \end{aligned}\right. \tag{6.122}$$

by (6.120). Choosing $f = \Phi_\pm^* h'$ in (6.122) gives (6.121).

Now let $0 < M < M'$ and define

$$\Psi(|p|^2) = e^{-it|p|} \chi_{(M,M')}(|p|), \quad t \in \mathbb{R}. \tag{6.123}$$

Then (6.121) and Lemma 6.17 imply that

$$\int_{M \leq |p| \leq M'} w_\pm(x,p) e^{-it|p|} h(p)\, dp = (\Phi_\pm^* h')(x) = 0 \tag{6.124}$$

in $L_2(\Omega)$ for all $t \in \mathbb{R}$ and $M, M' \in \mathbb{R}_+$ with $M' > M$.

If the decomposition (6.25) for $w_{\pm}(x,p)$, proved in Corollary 6.2, is substituted in (6.124) the equation can be rewritten

$$(6.125)\quad \begin{cases} j(x)u_0(t,x) + u_1(t,x) + u_2(t,x) = 0 \\ \text{in } L_2(\Omega) \quad \text{for } t \in \mathbb{R} \end{cases}$$

where

$$(6.126)\quad u_0(t,x) = \frac{1}{(2\pi)^{n/2}} \int_{M\le|p|\le M'} e^{i(x\cdot p - t|p|)} h(p)\,dp,$$

$$(6.127)\quad \begin{cases} u_1(t,x) \\ = \dfrac{1}{|x|^{\frac{n-1}{2}}} \displaystyle\int_{M\le|p|\le M'} e^{i(\pm|x|-t)|p|}\,\theta_{\pm}\left(\frac{x}{|x|}, p\right) h(p)\,dp \end{cases}$$

and

$$(6.128)\quad u_2(t,x) = \int_{M\le|p|\le M'} q_{\pm}(x,p)\, e^{-it|p|} h(p)\,dp.$$

Note that

$$(6.129)\qquad u_0(t,\cdot) = e^{-itA_0^{\frac{1}{2}}} h^0_{M,M'} \in L_2(\mathbb{R}^n)$$

where

$$(6.130)\qquad h^0_{M,M'} = \Phi^* \chi_{M,M'}(|\cdot|) h$$

and Φ is the Fourier transform in $L_2(\mathbb{R}^n)$. Thus

$$(6.131)\quad \begin{cases} \|u_0(t,\cdot)\|_{L_2(\mathbb{R}^n)} = \|h^0_{M,M'}\|_{L_2(\mathbb{R}^n)} \\[2ex] = \|\chi_{M,M'}(|\cdot|)h\|_{L_2(\mathbb{R}^n)} = \displaystyle\int_{M\le|p|\le M'} |h(p)|^2 dp. \end{cases}$$

To complete the proof of Theorem 6.14 it will be shown that

$$(6.132)\qquad \lim_{t\to\mp\infty} \|u_0(t,\cdot)\|_{L_2(\mathbb{R}^n)} = 0.$$

Combining (6.131) and (6.132) gives

$$(6.133)\qquad \int_{M\le|p|\le M'} |h(p)|^2 dp = 0 \quad \text{for all} \quad 0 < M < M'$$

which clearly implies $h = 0$ in $L_2(\mathbb{R}^n)$ and hence (6.112).

To prove (6.132) note first that

$$(6.134)\qquad u_1(t,x) = \frac{F_\pm(\pm|x|-t, x/|x|)}{|x|^{\frac{n-1}{2}}}$$

where

$$(6.135)\quad \begin{cases} F_\pm(r,\eta) = \displaystyle\int_{M\le|p|\le M'} e^{ir|p|}\theta_\pm(\eta,p)h(p)\,dp, \\[2ex] r \in \mathbb{R}, \quad \eta \in S^{n-1}. \end{cases}$$

Moreover, $F_\pm \in L_2(\mathbb{R} \times S^{n-1})$. To verify this note that introducing spherical coordinates $p = \rho\omega$, $\rho \ge 0$, $\omega \in S^{n-1}$ in (6.135) gives

$$(6.136)\quad F_{\pm}(r,\eta) = \int_M^{M'} e^{ir\rho}\left(\int_{S^{n-1}} \theta_{\pm}(\eta,\rho\omega)h(\rho\omega)\rho^{n-1}d\omega\right)d\rho .$$

Recall that $\theta_{\pm}(\eta,p) \in C^{\infty}(S^{n-1} \times \mathbb{R}^n - \{0\})$, by Corollary 6.2. An application of Parseval's formula and Fubini's theorem gives

$$(6.137)\quad \begin{cases} \|F_{\pm}\|^2_{L_2(\mathbb{R}\times S^{n-1})} = 2\pi \int_M^{M'} \int_{S^{n-1}} \\ \left(\int_{S^{n-1}} \left|\theta_{\pm}(\eta,\rho\omega)\rho^{\frac{n-1}{2}}\right|^2 d\eta\right)\left|h(\rho\omega)\rho^{\frac{n-1}{2}}\right|^2 d\omega d\rho . \end{cases}$$

Now $\left|\theta_{\pm}(\eta,p)|p|^{\frac{n-1}{2}}\right| \le C$ for $\eta \in S^{n-1}$ and $M \le |p| \le M'$ by Corollary 6.2 where $C = C(M,M')$ is a suitable constant. Thus (6.137) implies

$$(6.138)\quad \|F_{\pm}\|^2_{L_2(\mathbb{R}\times S^{n-1})} \le 2\pi |S^{n-1}| C^2 \|h\|^2_{L_2(\mathbb{R}^n)} < \infty .$$

The representation (6.134), together with (6.138) and Theorem 2.5 imply that

$$(6.139)\quad \lim_{t\to\mp\infty} \|u_1(t,\cdot)\|_{L_2(\mathbb{R}^n)} = 0 .$$

Finally, Lemma 2.7 will be used to prove that

$$(6.140)\quad \lim_{t\to\mp\infty} \|u_2(t,\cdot)\|_{L_2(\Omega)} = 0 .$$

To verify the hypotheses of the lemma, note first that (6.125) implies that $u_2(t,\cdot) \in L_2(\Omega)$ because $u_0(t,\cdot) \in$

$L_2(\mathbb{R}^n)$ and $u_1(t,\cdot) \in L_2(\mathbb{R}^n)$. Moreover, (6.125) also implies

(6.141) $$\lim_{t \to \mp\infty} \|u_2(t,\cdot)\|_{L_2(\Omega_r)} = 0$$

for each $r > 0$, because $u_0(t,\cdot)$ has this property by Theorem 5.5 and

(6.142) $$\|u_1(t,\cdot)\|^2_{L_2(\Omega_r)} = \int_{-t}^{r-t} \int_{S^{n-1}} |F(r',\eta)|^2 d\eta dr',$$

which tends to zero when $|t| \to \infty$ by (6.138). Thus $u_2(t,x)$ satisfies (2.69), (2.70). Finally the estimate (6.26) for $q_\pm(x,p)$ of Corollary 6.2 implies

(6.143) $$|u_2(t,x)| \le \int_{M\le|p|\le M'} |h(p)|dp/|x|^{\frac{n+1}{2}} = \frac{C'(M,M',h)}{|x|^{\frac{n+1}{2}}}$$

for all $t \in \mathbb{R}$ and $|x| \ge r_0$ which verifies (2.71) for $u_2(t,x)$. This completes the verification of (6.141).

Finally, to prove (6.132) note that if $r > r_0 + 1$ then (6.125) implies

(6.144) $$\begin{cases} \|u_0(t,\cdot)\|_{L_2(\mathbb{R}^n)} \le \|u_0(t,\cdot)\|_{L_2(B(r))} \\ \quad + \|u_0(t,\cdot)\|_{L_2(\mathbb{R}^n - B(r))} \le \|u_0(t,\cdot)\|_{L_2(B(r))} \\ \quad + \|u_1(t,\cdot)\|_{L_2(\Omega)} + \|u_2(t,\cdot)\|_{L_2(\Omega)} . \end{cases}$$

Now the three terms on the right tend to zero when $t \to \mp\infty$, by Theorem 5.5, (6.139) and (6.141). This completes the proof of (6.132) and Theorem 6.14.

Corollary 6.13 and Theorem 6.15 imply that the two eigenfunction expansions based on $w_\pm(x,p)$ provide spectral representations of the operator A. This result provides the basis for the study in Lectures 7 and 8 of solutions of the d'Alembert equation in Ω. It may be formulated as follows:

COROLLARY 6.18. *If* $\Psi(\lambda)$ *is a bounded, Lebesgue measurable function of* $\lambda \geq 0$ *then* $\Psi(A)$ *has the representation*

$$(6.145) \quad \Psi(A)f(x) = \underset{M\to\infty}{L_2(\Omega)\text{-}\lim} \int_{|p|\leq M} w_\pm(x,p)\Psi(|p|^2)\hat{f}_\pm(p)\,dp.$$

PROOF. (6.145) follows immediately from (6.97) and (6.103).

LECTURE 7. WAVE OPERATORS AND ASYMPTOTIC SOLUTIONS OF THE D'ALEMBERT EQUATION IN EXTERIOR DOMAINS

In this lecture the solution in $L_2(\Omega)$ of the initial-boundary value problem is constructed by means of the eigenfunction expansions of Lecture 6 and its asymptotic behavior for $t \to \infty$ is calculated. The principal result of the lecture is the theorem that every solution in $L_2(\Omega)$ is asymptotically equal to a free wave in $L_2(\mathbb{R}^n)$, as was suggested in Lecture 5. As a corollary, the existence of the wave operators $W_\pm$ is proved and an explicit representation for them is derived. The results exhibit the precise relationship between the steady-state scattering theory of Lectures 4 and 6 and the time-dependent scattering theory of Lecture 5.

The complex-valued solutions in $L_2(\Omega)$ defined by

$$(7.1) \qquad v(t,x) = e^{-itA^{\frac{1}{2}}} h(x), \quad h \in L_2(\Omega)$$

are studied in this lecture. Corresponding results for real-valued solutions are given in Lecture 8. Corollary 6.18 implies that $v(t,x)$ has two spectral representations corresponding to the two complete sets of eigenfunctions $\{w_+(x,p) : p \in \mathbb{R}^n\}$ and $\{w_-(x,p) : p \in \mathbb{R}^n\}$. They have the form

$$(7.2) \qquad v(t,x) = \underset{M\to\infty}{L_2(\Omega) - \lim} \int_{|p|\le M} w_+(x,p)\, e^{-it|p|} \hat{h}_+(p)\, dp$$

and

$$(7.3)\qquad v(t,x) = L_2(\Omega)-\lim_{M\to\infty} \int_{|p|\le M} w_-(x,p)\, e^{-it|p|}\hat{h}_-(p)\,dp,$$

where

$$(7.4)\qquad \hat{h}_\pm(p) = L_2(\mathbb{R}^n)-\lim_{M\to\infty} \int_{\Omega_M} \overline{w_\pm(x,p)}\, h(x)\,dx = \Phi_\pm h(p).$$

Recall that

$$(7.5)\qquad w_\pm(x,p) = j(x)\, w_0(x,p) + w_\pm'(x,p),$$

where $w_+'(x,p)$ and $w_-'(x,p)$ behave like outgoing and incoming waves, respectively. For this reason (7.2) and (7.3) will be called the outgoing and incoming representations, respectively.

In principle, either of the representations (7.2), (7.3) can be used to calculate the asymptotic behavior of $v(t,x)$ for $t \to +\infty$. It will be shown that the incoming representation (7.3) leads to a particularly simple result. Substituting the decomposition (7.5) for $w_-(x,p)$ in (7.3) gives

$$(7.6)\qquad v(t,x) = j(x)\, v_0^+(t,x) + v^+(t,x)$$

where

$$(7.7)\qquad v_0^+(t,x) = L_2(\mathbb{R}^n)-\lim_{M\to\infty} \int_{|p|\le M} w_0(x,p)\, e^{-it|p|}\, \hat{h}_-(p)\,dp$$

and

$$(7.8) \quad v^+(t,x) = \underset{M\to\infty}{L_2(\Omega) - \lim} \int_{|p|\le M} w_-'(x,p)\, e^{-it|p|} \hat{h}_-(p)\,dp.$$

In particular, the existence of the limits in (7.3) and (7.7), which is known from the eigenfunction expansion theorems for A and A_0, implies that the limit in (7.8) exists. Note that $v_0^+(t,x)$ is a free wave in $L_2(\mathbb{R}^n)$ of the form

$$(7.9) \quad v_0^+(t,x) = e^{-itA_0^{1/2}} h_0^+(x),$$

where $h_0^+ \in L_2(\mathbb{R}^n)$ is defined by $h_0^+(x) = v_0^+(0,x)$. Thus, by (6.4) and (7.4)

$$(7.10) \quad h_0^+ = \Phi^* \hat{h}_- = \Phi^* \Phi_- h,$$

where Φ is the Fourier transform in $L_2(\mathbb{R}^n)$. With this notation the main results of this lecture may be stated as follows.

THEOREM 7.1. *Let* Ω *be an exterior domain such that* $\Omega \in LC$. *Then for every* $h \in L_2(\Omega)$

$$(7.11) \quad \underset{t\to+\infty}{L_2(\Omega) - \lim} v^+(t,\cdot) = 0,$$

and hence

$$(7.12) \quad \lim_{t\to+\infty} \| v(t,\cdot) - j(\cdot) v_0^+(t,\cdot) \|_{L_2(\Omega)} = 0.$$

COROLLARY 7.2. *If* Ω *is an exterior domain such that* $\Omega \in LC$ *then the wave operator* $W_+ = W_+(A_0^{1/2}, A^{1/2}, J_\Omega)$ *exists and*

(7.13) $$W_+ = \Phi^* \Phi_- .$$

In particular, $W_+ : L_2(\Omega) \to L_2(\mathbb{R}^n)$ *is unitary. Moreover, it defines a unitary equivalence between* A *and* A_0*:*

(7.14) $$\Pi(\lambda) = W_+^* \Pi_0(\lambda) W_+ \quad \text{for all} \quad \lambda \in \mathbb{R}.$$

Note that Theorem 5.6 of Lecture 5 is proved by Corollary 7.2.

PROOF OF THEOREM 7.1. The statements (7.11) and (7.12) are equivalent, by (7.6). Moreover, (7.12) can be written

(7.15) $$\lim_{t \to +\infty} \| (e^{-itA^{\frac{1}{2}}} - J^* e^{-itA_0^{\frac{1}{2}}} \Phi^* \Phi_-) h \|_{L_2(\Omega)} = 0$$

by (7.1), (7.9) and (7.10) where $J^* : L_2(\mathbb{R}^n) \to L_2(\Omega)$ is defined by

(7.16) $$J^* f(x) = j(x) f(x) \quad \text{for all} \quad x \in \Omega.$$

The notation J^* is used because (7.16) defines the adjoint of the operator $J: L_2(\Omega) \to L_2(\mathbb{R}^n)$ defined by (6.34). Now the operators $e^{-itA^{\frac{1}{2}}} - J^* e^{-itA_0^{\frac{1}{2}}} \Phi^* \Phi_-$ are uniformly bounded for all $t \in \mathbb{R}$. It follows that to prove Theorem 7.1 it is enough to verify (7.15) for the vectors h in a dense subset of $L_2(\Omega)$, as in the proofs of Theorems 2.6 and 5.3. It will be convenient to use the dense set

(7.17) $\mathcal{D}_0^- = \Phi_-^* \mathcal{D}_0(\mathbb{R}^n) = \{h \in L_2(\Omega): \hat{h}_- \in \mathcal{D}_0(\mathbb{R}^n)\}$

where $\mathcal{D}_0(\mathbb{R}^n)$ is the set defined in Lecture 2, (2.36). $\mathcal{D}_0^-$ is dense in $L_2(\Omega)$ because $\mathcal{D}_0(\mathbb{R}^n)$ is dense in $L_2(\mathbb{R}^n)$ and $\Phi_-^*: L_2(\mathbb{R}^n) \to L_2(\Omega)$ is unitary, by Theorem 6.14.

Assume that $\hat{h}_- \in \mathcal{D}_0(\mathbb{R}^n)$ so that

$$(7.18) \qquad v^+(t,x) = \int_K w_-'(x,p)\, e^{-it|p|} \hat{h}_-(p)\, dp$$

where K, the support of $\hat{h}_-$, is compact. The convergence (7.11) will be proved by means of Lemma 2.7. Recall that $w_-'(x,p)$ is a solution of

$$(7.19) \qquad (\Delta + |p|^2) w_-'(x,p) = M(x,p), \quad x \in \Omega, \quad p \in \mathbb{R}^n$$

where $M(x,p) \in C^\infty(\mathbb{R}^n \times \mathbb{R}^n)$. In particular, $w_-'(x,p) \in C^\infty(\Omega \times \mathbb{R}^n)$. Moreover, Corollary 6.2 implies that

$$(7.20) \qquad w_-'(x,p) = \frac{e^{-i|p||x|}}{|x|^{\frac{n-1}{2}}} \theta_-\left(\frac{x}{|x|}, p\right) + q_-(x,p)$$

where $\theta_-(\eta,p) \in C^\infty(S^{n-1} \times \mathbb{R}^n - \{0\})$ and for each compact $K \subset \mathbb{R}^n - \{0\}$ there is a constant $M_- = M_-(K)$ such that

$$(7.21) \qquad \begin{cases} |q_-(x,p)| \le \dfrac{M_-}{|x|^{\frac{n+1}{2}}} & \text{for all } x \in \Omega - \{0\} \\ \text{and } p \in K. \end{cases}$$

Substituting (7.20) into (7.18) gives

(7.22) $$v^{+}(t,x) = v_1^{+}(t,x) + v_2^{+}(t,x)$$

where

(7.23) $$v_1^{+}(t,x) = \frac{1}{|x|^{\frac{n-1}{2}}} \int_K e^{-i|p|(|x|+t)}\, \theta_-\!\left(\frac{x}{|x|}, p\right) \hat{h}_-(p)\,dp$$

and

(7.24) $$v_2^{+}(t,x) = \int_K q_-(x,p)\, e^{-it|p|}\, \hat{h}_-(p)\,dp.$$

Introduction of spherical coordinates $p = \rho\omega$ in (7.23) gives

(7.25) $$\begin{cases} v_1^{+}(t,x) = \dfrac{1}{|x|^{\frac{n-1}{2}}} \displaystyle\int_a^b e^{-i(|x|+t)\rho} \\ \cdot \left\{ \displaystyle\int_{S^{n-1}} \theta_-\!\left(\frac{x}{|x|}, \rho\omega\right) \hat{h}_-(\rho\omega)\, \rho^{n-1} d\omega \right\} d\rho \end{cases}$$

where a and b are numbers such that $K = \operatorname{supp} \hat{h}_- \subset \{p\colon 0 < a \leq |p| \leq b\}$. Moreover, the function in braces in (7.25) is in the class $\mathcal{D}(\mathbb{R}_+)$ as a function of ρ. Hence, an integration by parts gives

(7.26) $$\begin{cases} v_1^{+}(t,x) = \dfrac{1}{|x|^{\frac{n-1}{2}}} \displaystyle\int_a^b \frac{e^{-i(|x|+t)\rho}}{i(|x|+t)} \\ \cdot \left\{ \dfrac{\partial}{\partial\rho} \displaystyle\int_{S^{n-1}} \theta_-\!\left(\frac{x}{|x|}, \rho\omega\right) \hat{h}_-(\rho\omega)\, \rho^{n-1} d\omega \right\} d\rho, \end{cases}$$

and the function in braces in (7.26) is bounded for all $a \le \rho \le b$ and $x/|x| \in S^{n-1}$. Let M_0 be a bound for it so that

$$(7.27)\quad |v_1^+(t,x)| \le \frac{1}{|x|^{\frac{n-1}{2}}(|x|+t)} \int_a^b M_0 d\rho \le \frac{M_0(b-a)}{|x|^{\frac{n+1}{2}}}$$

for all $x \in \Omega - \{0\}$ and $t > 0$. The last inequality holds because $|x| + t > |x|$ for $t > 0$. Combining (7.21) and (7.24) gives a similar estimate for $v_2^+(t,x)$, namely

$$(7.28)\quad |v_2^+(t,x)| \le \int_K |q_-(x,p)|\,|\hat{h}_-(p)|\,dp \le \frac{M_- M_1}{|x|^{\frac{n+1}{2}}}$$

for all $x \in \Omega - \{0\}$ and $t \in \mathbb{R}$ where $M_1 = \int_K |\hat{h}_-(p)|\,dp$. Combining (7.22), (7.27) and (7.28) gives

$$(7.29)\quad |v^+(t,x)| \le \frac{N}{|x|^{\frac{n+1}{2}}} \quad \text{for all } t>0 \text{ and } x \in \Omega - \{0\}$$

where $N = N(h)$ is a suitable constant.

Equation (7.29) is condition (2.71) of Lemma 2.7. Equation (7.6) implies that $v^+(t,\cdot) \in L_2(\Omega)$ for all $t \in \mathbb{R}$ because $v(t,\cdot) \in L_2(\Omega)$ and $v_0^+(t,\cdot) \in L_2(\mathbb{R}^n)$. Finally $v^+(t,x)$ satisfies the local decay condition

$$(7.30)\quad \begin{cases} \lim_{t\to\infty} \|v^+(t,\cdot)\|_{L_2(K\cap\Omega)} = 0 \\ \text{for every compact } K \subset \mathbb{R}^n \end{cases}$$

because both $v(t,x)$ and $v_0^+(t,x)$ have this property

by Theorem 5.5. Thus Lemma 2.7 implies (7.11) when $h \in \mathcal{D}_0^-$. The extension to the case of arbitrary $h \in L_2(\Omega)$ is done by using the density of $\mathcal{D}_0^-$ in $L_2(\Omega)$. The argument will not be repeated here.

PROOF OF COROLLARY 7.2. The starting point for the proof is equation (7.15) which holds for all $h \in L_2(\Omega)$ by Theorem 7.1. Note that if $J_\Omega: L_2(\Omega) \to L_2(\mathbb{R}^n)$ and $J: L_2(\Omega) \to L_2(\mathbb{R}^n)$ are defined by (5.21) and (6.34), respectively, then

$$(7.31)\quad \begin{cases} J_\Omega e^{-itA^{\frac{1}{2}}} - e^{-itA_0^{\frac{1}{2}}} \Phi^* \Phi_- \\ = Je^{-itA^{\frac{1}{2}}} - e^{-itA_0^{\frac{1}{2}}} \Phi^* \Phi_- + (J_\Omega - J) e^{-itA^{\frac{1}{2}}} \\ = J(e^{-itA^{\frac{1}{2}}} - J^* e^{-itA_0^{\frac{1}{2}}} \Phi^* \Phi_-) \\ \quad + (JJ^* - 1) e^{-itA_0^{\frac{1}{2}}} \Phi^* \Phi_- + (J_\Omega - J) e^{-itA^{\frac{1}{2}}}. \end{cases}$$

Note that $(JJ^* - 1) f(x) = (j^2(x) - 1) f(x)$ for all $x \in \mathbb{R}^n$ and

$$(7.32)\quad (J_\Omega - J) f(x) = \begin{cases} (1 - j(x)) f(x), & x \in \Omega \\ 0 & , x \in \mathbb{R}^n - \Omega. \end{cases}$$

In particular $\|JJ^* - 1\| \le 1$, $\|J_\Omega - J\| \le 1$, $\operatorname{supp}(JJ^* - 1) f \subset \{x: |x| \le r_0\}$ and $\operatorname{supp}(J_\Omega - J) f \subset \{x: |x| \le r_0\}$. Thus (7.31) implies that, for all $h \in L_2(\Omega)$,

$$(7.33)\quad \begin{cases} \|(J_\Omega e^{-itA^{\frac12}} - e^{-itA_0^{\frac12}}\Phi^*\Phi_-)h\|_{L_2(\mathbb{R}^n)} \\ \le \|(e^{-itA^{\frac12}} - J^* e^{-itA_0^{\frac12}}\Phi^*\Phi_-)h\|_{L_2(\Omega)} \\ + \|e^{-itA_0^{\frac12}}(\Phi^*\Phi_- h)\|_{L_2(B(r_0))} + \|e^{-itA^{\frac12}} h\|_{L_2(\Omega_{r_0})} . \end{cases}$$

Now the first term on the right in (7.33) tends to zero when $t \to +\infty$ by (7.15). Moreover, the last two terms tend to zero when $t \to +\infty$ by Theorem 5.5, applied to A_0 and A respectively. Thus (7.33) implies

$$(7.34)\quad \lim_{t\to+\infty} \|(J_\Omega e^{-itA^{\frac12}} - e^{-itA_0^{\frac12}}\Phi^*\Phi_-)h\|_{L_2(\mathbb{R}^n)} = 0$$

for all $h \in L_2(\Omega)$. This is equivalent to the equation

$$(7.35)\quad \lim_{t\to+\infty} \|e^{itA_0^{\frac12}} J_\Omega e^{-itA^{\frac12}} h - \Phi^*\Phi_- h\|_{L_2(\mathbb{R}^n)} = 0$$

for all $h \in L_2(\Omega)$, because $e^{itA_0^{\frac12}}$ is unitary. Equation (7.35) implies that $W_+ = W_+(A_0^{\frac12}, A^{\frac12}, J_\Omega)$ exists and satisfies (7.13).

The unitarity of $W_+ : L_2(\Omega) \to L_2(\mathbb{R}^n)$ follows from the unitarity of $\Phi_- : L_2(\Omega) \to L_2(\mathbb{R}^n)$ (Theorem 6.14) and $\Phi^* : L_2(\mathbb{R}^n) \to L_2(\mathbb{R}^n)$. To complete the proof of Corollary 7.2 it remains to verify (7.14). Note that (7.13) and the unitarity of Φ and Φ_- imply that $W_+^* = \Phi_-^*\Phi = W_+^{-1}$. Hence, (7.14) can be written in the equivalent forms

$$(7.36)\qquad \Pi(\lambda) = \Phi_-^*\Phi\Pi_0(\lambda)\Phi^*\Phi_- \quad \text{for all } \lambda \in \mathbb{R}$$

or

$$(7.37) \qquad \Phi_- \Pi(\lambda) \Phi_-^* = \Phi \Pi_0(\lambda) \Phi^* \quad \text{for all} \quad \lambda \in \mathbb{R}.$$

But the last equation is correct because the two sides coincide with the operation of multiplying with $H(\lambda - |p|^2)$ in $L_2(\mathbb{R}^n)$, by Theorem 6.15 and equation (2.24).

Now consider the outgoing representation of $v(t,x)$, equation (7.2). Substituting the decomposition (7.5) for $w_+(x,p)$ in (7.2) gives

$$(7.38) \qquad v(t,x) = j(x) v_0^-(t,x) + v^-(t,x)$$

where

$$(7.39) \quad v_0^-(t,x) = \underset{M \to \infty}{L_2(\mathbb{R}^n) - \lim} \int_{|p| \le M} w_0(x,p) e^{-it|p|} \hat{h}_+(p)\, dp$$

and

$$(7.40) \quad v^-(t,x) = \underset{M \to \infty}{L_2(\Omega) - \lim} \int_{|p| \le M} w_+'(x,p) e^{-it|p|} \hat{h}_+(p)\, dp.$$

Moreover, $v_0^-(t,x)$ is a free wave in $L_2(\mathbb{R}^n)$ of the form

$$(7.41) \qquad v_0^-(t,x) = e^{-itA_0^{1/2}} h_0^-(x)$$

where

$$(7.42) \qquad h_0^- = \Phi^* \hat{h}_+ = \Phi^* \Phi_+ h.$$

Also, substituting the representation

$$(7.43)\qquad w_+'(x,p) = \frac{e^{i|p||x|}}{|x|^{\frac{n-1}{2}}}\,\theta_+\left(\frac{x}{|x|},p\right) + q_+(x,p)$$

from Corollary 6.2 into (7.40) gives

$$(7.44)\qquad v^-(t,x) = v_1^-(t,x) + v_2^-(t,x)$$

where

$$(7.45)\quad v_1^-(t,x) = \frac{1}{|x|^{\frac{n-1}{2}}}\int_K e^{i|p|(|x|-t)}\,\theta_+\left(\frac{x}{|x|},p\right)\hat{h}_+(p)\,dp$$

and

$$(7.46)\qquad v_2^-(t,x) = \int_K q_+(x,p)\,e^{-it|p|}\,\hat{h}_+(p)\,dp.$$

In the last two equations it is assumed, for simplicity, that $\hat{h}_+ \in \mathcal{D}_0(\mathbb{R}^n)$ and $\operatorname{supp}\hat{h}_+ = K$. The estimate (6.26) for $q_+(x,p)$ implies that $|v_2^-(t,x)| \le M/|x|^{\frac{n+1}{2}}$ for all $t \in \mathbb{R}$ and hence $\|v_2^-(t,\cdot)\| \to 0$ when $t \to \pm\infty$, by the argument used to prove Theorem 7.1. However, integration by parts in (7.45) does not lead to a similar result for $v_1^-(t,x)$ when $t \to +\infty$. This is because the term $|x| - t$ in the denominator is not larger than $|x|$ for $t > 0$. However, $|x| - t > |x|$ for all $t < 0$ and hence the method used to prove Theorem 7.1 shows that

$$(7.47)\qquad L_2(\Omega) - \lim_{t\to-\infty} v^-(t,\cdot) = 0.$$

Thus the outgoing representation leads to the following

analogues of Theorem 7.1 and Corollary 7.2.

COROLLARY 7.3. *Let* Ω *be an exterior domain such that* $\Omega \in LC$. *Then for every* $h \in L_2(\Omega)$

$$(7.48) \qquad \lim_{t\to-\infty} \|v(t,\cdot) - j(x)v_0^-(t,\cdot)\|_{L_2(\Omega)} = 0.$$

COROLLARY 7.4. *Under the same hypothesis the wave operator*

$$(7.49) \qquad W_- = W_-(A_0^{\frac{1}{2}}, A^{\frac{1}{2}}, J_\Omega) = \text{s-}\lim_{t\to-\infty} e^{itA_0^{\frac{1}{2}}} J_\Omega e^{-itA^{\frac{1}{2}}}$$

exists and

$$(7.50) \qquad W_- = \Phi^* \Phi_+ .$$

In particular, $W_- : L_2(\Omega) \to L_2(\mathbb{R}^n)$ *is unitary and satisfies*

$$(7.51) \qquad \Pi(\lambda) = W_-^* \Pi_0(\lambda) W_- \quad \textit{for all} \quad \lambda \in \mathbb{R}.$$

It should be noted that the proof of Theorem 7.1 and the corollaries depends in an essential way on Theorem 5.5 which guarantees the "local decay" property of solutions in $L_2(\Omega)$. In Lecture 8, Theorem 7.1 and the results on asymptotic wave functions of Lecture 2 are combined to obtain stronger results on the decay of solutions in $L_2(\Omega)$ and solutions wFE. The remainder of this lecture is devoted to showing that if $v(t,x)$ is a solution wFE then the asymptotic solutions $v_0^+(t,x)$ and $v_0^-(t,x)$ of Theorem 7.1 and Corollary 7.3 approximate $v(t,x)$ in the energy norm. More precisely, the following result is proved.

THEOREM 7.5. *Let* Ω *be an exterior domain such that* $\Omega \in LC$ *and let* $h \in D(A^{\frac{1}{2}}) = L_2^1(\Omega)$ *so that* $v(t,\cdot) = e^{-itA^{\frac{1}{2}}} h$ *is a solution* wFE. *Then* $h_0^{\pm} = W_{\pm} h \in D(A_0^{\frac{1}{2}}) = L_2^1(\mathbb{R}^n)$, *so that* $v_0^{\pm}(t,\cdot) = e^{-itA_0^{\frac{1}{2}}} h_0^{\pm}$ *are solutions* wFE, *and*

$$\lim_{t\to\pm\infty} \| D_k v(t,\cdot) - j(\cdot) D_k v_0^{\pm}(t,\cdot) \|_{L_2(\Omega)} = 0 \tag{7.52}$$

for $k = 0,1,2,\ldots,n$.

PROOF. Only the limit $t \to +\infty$ will be discussed. The other case is exactly analogous. Note that Corollary 7.2 implies that $h \in D(A^{\frac{1}{2}})$ if and only if $h_0^+ = W_+ h \in D(A_0^{\frac{1}{2}})$. To prove (7.52) with $k = 0$ note that when $h \in D(A^{\frac{1}{2}})$ then

$$D_0 v(t,\cdot) = -iA^{\frac{1}{2}} e^{-itA^{\frac{1}{2}}} h = -ie^{-itA^{\frac{1}{2}}} (A^{\frac{1}{2}} h) \tag{7.53}$$

and

$$\begin{cases} D_0 v_0^+(t,\cdot) = -iA_0^{\frac{1}{2}} e^{-itA_0^{\frac{1}{2}}} h_0^+ = -ie^{-itA_0^{\frac{1}{2}}} A_0^{\frac{1}{2}} W_+ h \\ \qquad = -ie^{-itA_0^{\frac{1}{2}}} W_+ (A^{\frac{1}{2}} h) . \end{cases} \tag{7.54}$$

The last step in (7.54) follows from

$$W_+ A^{\frac{1}{2}} \subset A_0^{\frac{1}{2}} W_+ \tag{7.55}$$

which is a consequence of (7.14). Now Theorem 7.1 states that

$$(7.56)\qquad \lim_{t\to+\infty} \| e^{-itA^{\frac{1}{2}}} h - J^* e^{-itA_0^{\frac{1}{2}}} W_+ h \|_{L_2(\Omega)} = 0$$

for all $h \in L_2(\Omega)$. Substituting $A^{\frac{1}{2}}h$ for h in (7.56) and using (7.53), (7.54) gives (7.52) with $k = 0$.

The proof of (7.52) with $k = 1,2,\ldots,n$ is more difficult. Note first that, since $jD_k v_0^+ = D_k(jv_0^+) + (D_k j)v_0^+$,

$$(7.57)\qquad \begin{cases} \|D_k v(t,\cdot) - j(\cdot)D_k v_0^+(t,\cdot)\| \\ \le \|D_k\{v(t,\cdot) - j(\cdot)v_0^+(t,\cdot)\}\| + \|D_k j(\cdot)v_0^+(t,\cdot)\|. \end{cases}$$

Now $\operatorname{supp} D_k j \subset B(r_0 + 1)$ and hence the last term tends to zero when $t \to \infty$ by Theorem 5.5. Thus, to prove (7.52) for $k = 1,2,\ldots,n$ it will suffice to prove that

$$(7.58)\qquad \lim_{t\to+\infty} \|\nabla\{v(t,\cdot) - j(\cdot)v_0^+(t,\cdot)\}\|_{L_2(\Omega)} = 0.$$

To prove this assume first that $h \in D(A)$. Then $h_0^+ = W_+h \in D(A_0)$, by (7.14), and it follows that $v(t,\cdot) \in D(A)$ and $v_0^+(t,\cdot) \in D(A_0)$ for all $t \in \mathbb{R}$. Moreover, $h \in D(A^{\frac{1}{2}}) = L_2^1(\Omega)$ and $h_0^+ = W_+h \in D(A_0^{\frac{1}{2}}) = L_2^1(\mathbb{R}^n)$. It follows that $j(x)v_0^+(t,x) \in L_2^1(\Omega) = D(A^{\frac{1}{2}})$ and hence

$$(7.59)\quad \begin{cases} \|\nabla\{v(t,\cdot)-j(\cdot)v_0^+(t,\cdot)\}\|^2 \\ = \|A^{\frac{1}{2}}\{v(t,\cdot)-j(\cdot)v_0^+(t,\cdot)\}\|^2 \\ = (A^{\frac{1}{2}}\{v(t,\cdot)-j(\cdot)v_0^+(t,\cdot)\},A^{\frac{1}{2}}\{v(t,\cdot)-j(\cdot)v_0^+(t,\cdot)\}) \\ = (A\{v(t,\cdot)-j(\cdot)v_0^+(t,\cdot)\},\{v(t,\cdot)-j(\cdot)v_0^+(t,\cdot)\}) \\ \leq \|Av(t,\cdot)-AJ^*v_0^+(t,\cdot)\| \ \|v(t,\cdot)-j(\cdot)v_0^+(t,\cdot)\| \\ \leq (\|Av(t,\cdot)\| + \|AJ^*v_0^+(t,\cdot)\|)\|v(t,\cdot)-j(\cdot)v_0^+(t,\cdot)\| . \end{cases}$$

Note that

$$(7.60)\quad \|Av(t,\cdot)\| = \|Ae^{-itA^{\frac{1}{2}}}h\| = \|e^{-itA^{\frac{1}{2}}}Ah\| = \|Ah\| < \infty$$

because $h \in D(A)$. Moreover,

$$(7.61)\quad \begin{cases} AJ^*v_0^+(t,x) = -\Delta\{j(x)v_0^+(t,x)\} \\ = -(\Delta j(x)\cdot v_0^+(t,x)+2\nabla j(x)\cdot\nabla v_0^+(t,x)+j(x)\Delta v_0^+(t,x)) . \end{cases}$$

Thus, if M is a bound for $|\Delta j(x)|$ and $|\nabla j(x)|$,

$$(7.62)\quad \begin{cases} \|AJ^*v_0^+(t,\cdot)\| \leq M\|e^{-itA_0^{\frac{1}{2}}}h_0^+\| \\ + 2M\|\nabla e^{-itA_0^{\frac{1}{2}}}h_0^+\| + \|A_0e^{-itA_0^{\frac{1}{2}}}h_0^+\| \\ = M\|h_0^+\| + 2M\|A_0^{\frac{1}{2}}e^{-itA_0^{\frac{1}{2}}}h_0^+\| + \|e^{-itA_0^{\frac{1}{2}}}A_0h_0^+\| \\ = M\|h_0^+\| + 2M\|A_0^{\frac{1}{2}}h_0\| + \|A_0h_0^+\| < \infty \end{cases}$$

because $h_0^+ = W_+ h \in D(A_0)$. Thus (7.59) implies that for all $h \in D(A)$

(7.63) $$\|\nabla\{v(t,\cdot)-j(\cdot)v_0^+(t,\cdot)\}\|^2 \le C\|v(t,\cdot)-j(\cdot)v_0^+(t,\cdot)\|$$

where $C = C(h)$ is independent of t. Making $t \to \infty$ in (7.63) and using Theorem 7.1 gives (7.58) for the case of $h \in D(A)$.

The proof of (7.58) in the general case of $h \in D(A^{\frac{1}{2}})$ will be completed by a density argument. Note that

$$\|\nabla\{e^{-itA^{\frac{1}{2}}}h - J^* e^{-itA_0^{\frac{1}{2}}} W_+ h\}\|$$

$$= \|\nabla\{v(t,\cdot)-j(\cdot)v_0^+(t,\cdot)\}\|$$

$$= \|\nabla v(t,\cdot)-j(\cdot)\nabla v_0^+(t,\cdot)-(\nabla j)v_0^+(t,\cdot)\|$$

$$\le \|\nabla v(t,\cdot)\| + \|j(\cdot)\nabla v_0^+(t,\cdot)\| + \|(\nabla j)v_0^+(t,\cdot)\|$$

(7.64) $$\le \|A^{\frac{1}{2}}e^{-itA^{\frac{1}{2}}}h\|_{L_2(\Omega)} + \|\nabla e^{-itA_0^{\frac{1}{2}}}W_+h\|_{L_2(\mathbb{R}^n)}$$

$$+ M\|e^{-itA_0^{\frac{1}{2}}}W_+h\|_{L_2(\mathbb{R}^n)} = \|e^{-itA^{\frac{1}{2}}}A^{\frac{1}{2}}h\|_{L_2(\Omega)}$$

$$+ \|A_0^{\frac{1}{2}}e^{-itA_0^{\frac{1}{2}}}W_+h\|_{L_2(\mathbb{R}^n)} + M\|W_+h\|_{L_2(\mathbb{R}^n)}$$

$$= \|A^{\frac{1}{2}}h\|_{L_2(\Omega)} + \|A_0^{\frac{1}{2}}W_+h\|_{L_2(\mathbb{R}^n)} + M\|h\|_{L_2(\Omega)}$$

$$= \|A^{\frac{1}{2}}h\|_{L_2(\Omega)} + \|W_+A^{\frac{1}{2}}h\|_{L_2(\mathbb{R}^n)} + M\|h\|_{L_2(\Omega)}$$

$$(7.64)\ (\text{cont.})\quad = 2\,\|A^{\frac{1}{2}}h\|_{L_2(\Omega)} + M\,\|h\|_{L_2(\Omega)}$$

$$= 2\,\|\nabla h\|_{L_2(\Omega)} + M\,\|h\|_{L_2(\Omega)} \le M_1\,\|h\|_{L_2^1(\Omega)}\,.$$

Here M is a bound for $|\nabla j(x)|$ and $M_1 = 2+M$. To complete the proof of (7.58) note that $D(A)$ is dense in $D(A^{\frac{1}{2}}) = L_2^1(\Omega)$, by the spectral theorem. Given any $h \in D(A^{\frac{1}{2}})$ let $\{h_m\}$ be a sequence such that $h_m \in D(A)$ for $m = 1,2,\ldots$ and $\lim_{m\to\infty} h_m = h$ in $L_2^1(\Omega)$. Then, by (7.64) and the triangle inequality,

$$(7.65)\quad \begin{cases} \|\nabla\{v(t,\cdot)-j(\cdot)v_0^+(t,\cdot)\}\| \\ = \|\nabla(e^{-itA^{\frac{1}{2}}}-J^*e^{-itA_0^{\frac{1}{2}}}W_+)h\| \\ \le \|\nabla(e^{-itA^{\frac{1}{2}}}-J^*e^{-itA_0^{\frac{1}{2}}}W_+)h_m\| \\ +\ \|\nabla(e^{-itA^{\frac{1}{2}}}-J^*e^{-itA_0^{\frac{1}{2}}}W_+)(h-h_m)\| \\ \le \|\nabla(e^{-itA^{\frac{1}{2}}}h_m-J^*e^{-itA_0^{\frac{1}{2}}}W_+h_m)\| + M_1\,\|h-h_m\|_{L_2^1(\Omega)}\,. \end{cases}$$

Make $t \to +\infty$ in (7.65) with m fixed. The first term on the right-hand side tends to zero because $h_m \in D(A)$. Thus

$$(7.66)\quad \overline{\lim_{t\to+\infty}}\,\|\nabla\{v(t,\cdot)-j(\cdot)v_0^+(t,\cdot)\}\| \le M_1\,\|h-h_m\|_{L_2^1(\Omega)}$$

for $m = 1,2,3,\ldots$. This implies (7.58) because the

left-hand side of (7.66) is independent of m and $h_m \to h$ in $L_2^1(\Omega)$.

LECTURE 8. ASYMPTOTIC WAVE FUNCTIONS AND ENERGY DISTRIBUTIONS IN EXTERIOR DOMAINS

In this lecture, the results obtained in the preceding lectures are used to construct asymptotic wave functions for solutions in exterior domains Ω. These are then applied to calculate the asymptotic distribution of energy in bounded and unbounded subsets of Ω.

Consider first the complex-valued solution in $L_2(\Omega)$

$$(8.1) \qquad v(t,x) = e^{-itA^{\frac{1}{2}}} h(x), \quad h \in L_2(\Omega).$$

Theorem 7.1 states that $v(t,x)$ is asymptotically equal in $L_2(\Omega)$ to the free wave

$$(8.2) \qquad v_0^+(t,x) = e^{-itA_0^{\frac{1}{2}}} h_0^+(x)$$

for $t \to +\infty$ where

$$(8.3) \qquad h_0^+ = W_+ h = \Phi^* \Phi_- h.$$

Equation (7.12) of Theorem 7.1 is equivalent to

$$(8.4) \qquad \lim_{t\to+\infty} \| v(t,\cdot) - v_0^+(t,\cdot) \|_{L_2(\Omega)} = 0.$$

To see this, note that $v_0^+(t,x) - j(x)v_0^+(t,x) = (1-j(x))v_0^+(t,x)$ and hence

$$(8.5) \qquad \| (1-j(\cdot))v_0^+(t,\cdot) \|_{L_2(\Omega)} \le \| v_0^+(t,\cdot) \|_{L_2(\Omega_{r_0+1})} .$$

The last quantity tends to zero when $t \to +\infty$ by Theorem 5.5. Next, recall that the free wave (8.2) is asymptotically equal in $L_2(\mathbb{R}^n)$ to an asymptotic wave function

(8.6) $$v_0^{+,\infty}(t,x) = |x|^{\frac{1-n}{2}} G(|x|-t, x/|x|),$$

where $G(r,\eta)$ is defined by

(8.7) $$\hat{G}(\rho,\eta) = (-i\rho)^{\frac{n-1}{2}} H(\rho)\hat{h}_0^{+}(\rho\eta), \quad \rho \in \mathbb{R}, \quad \eta \in S^{n-1},$$

and $H(\tau)$ is the Heaviside function; see (2.57). By (8.3),

(8.8) $$\hat{h}_0^{+} = \Phi h_0^{+} = \Phi_- h = \hat{h}_- .$$

Thus (8.7) can also be written

(8.9) $$\hat{G}(\rho,\eta) = (-i\rho)^{\frac{n-1}{2}} H(\rho)\hat{h}_-(\rho\eta), \quad \rho \in \mathbb{R}, \quad \eta \in S^{n-1}.$$

Theorem 2.6 implies that

(8.10) $$\lim_{t\to+\infty} \| v_0^{+}(t,\cdot) - v_0^{+,\infty}(t,\cdot) \|_{L_2(\mathbb{R}^n)} = 0.$$

Now

(8.11) $$\left\{ \begin{aligned} &\| v(t,\cdot) - v_0^{+,\infty}(t,\cdot) \|_{L_2(\Omega)} \\ &\le \| v(t,\cdot) - v_0^{+}(t,\cdot) \|_{L_2(\Omega)} \\ &+ \| v_0^{+}(t,\cdot) - v_0^{+,\infty}(t,\cdot) \|_{L_2(\mathbb{R}^n)} \end{aligned} \right.$$

and the last two quantities tend to zero when $t \to +\infty$ by

(8.4) and (8.10). This suggests the

DEFINITION. The *asymptotic wave function* associated with the wave function (8.1) is

$$(8.12) \quad v^{\infty}(t,x) = |x|^{\frac{1-n}{2}} G(|x|-t, x/|x|), \; x \in \Omega - \{0\}, \; t \in \mathbb{R}$$

where $G \in L_2(\mathbb{R} \times S^{n-1})$ is defined by (8.9).

The basic properties of these asymptotic wave functions on Ω are given by the following analogue of Theorem 2.5.

THEOREM 8.1. *For all* $h \in L_2(\Omega)$ *the asymptotic wave function* (8.12) *satisfies*

$$(8.13) \qquad v^{\infty}(t,\cdot) \in L_2(\Omega) \text{ for all } t \in \mathbb{R},$$

$$(8.14) \quad t \to v^{\infty}(t,\cdot) \in L_2(\Omega) \text{ is continuous for all } t \in \mathbb{R},$$

$$(8.15) \qquad \lim_{t\to+\infty} \|v^{\infty}(t,\cdot)\|_{L_2(\Omega)} = \|h\|_{L_2(\Omega)},$$

$$(8.16) \qquad \lim_{t\to-\infty} \|v^{\infty}(t,\cdot)\|_{L_2(\Omega)} = 0.$$

PROOF. Properties (8.13), (8.14) and (8.16) follow from Theorem 2.5 because $G \in L_2(\mathbb{R} \times S^{n-1})$ and $\Omega \subset \mathbb{R}^n$. To verify (8.15) note that

$$(8.17) \qquad \lim_{t\to+\infty} \|v^{\infty}(t,\cdot)\|_{L_2(\Omega)} = \|G\|_{L_2(\mathbb{R}\times S^{n-1})}$$

by the argument used to prove Theorem 2.5. Now, by (8.9) and Theorem 6.14

$$(8.18)\quad \begin{cases} \|G\|^2_{L_2(\mathbb{R}\times S^{n-1})} = \|\hat{G}\|^2_{L_2(\mathbb{R}\times S^{n-1})} \\ = \int_0^\infty \int_{S^{n-1}} |\hat{h}_-(\rho\eta)|^2 \rho^{n-1} d\eta d\rho = \|\hat{h}_-\|^2_{L_2(\mathbb{R}^n)} \\ = \|h\|^2_{L_2(\Omega)} . \end{cases}$$

Combining (8.17) and (8.18) gives (8.15).

The basic convergence theorem for the asymptotic wave functions (8.12) is

THEOREM 8.2. *Let* Ω *be an exterior domain such that* $\Omega \in LC$. *Then for all* $h \in L_2(\Omega)$

$$(8.19)\qquad \lim_{t\to+\infty} \|v(t,\cdot) - v^\infty(t,\cdot)\|_{L_2(\Omega)} = 0.$$

PROOF. The hypothesis $\Omega \in LC$ is needed to guarantee the validity of the eigenfunction expansion theorem and the existence of the wave profile (8.9). Equation (8.19) follows from (8.11), (8.4) and (8.10), since v^∞ coincides with $v_0^{+,\infty}$ on Ω.

The asymptotic wave functions for solutions of the initial-boundary value problem for the d'Alembert equation in Ω follow immediately from Theorem 8.2. Let

$$(8.20)\qquad u(t,x) = (\cos tA^{\frac{1}{2}})f(x) + (A^{-\frac{1}{2}} \sin tA^{\frac{1}{2}})g(x)$$

be the solution in $L_2(\Omega)$ with initial values f and g.

Then Theorems 3.4 and 8.2 imply

COROLLARY 8.3. *Let* Ω *be an exterior domain such that* $\Omega \in LC$. *Let* f *and* g *be real-valued functions such that* $f \in L_2(\Omega)$ *and* $g \in D(A^{-\frac{1}{2}})$ *and define* $h = f + iA^{-\frac{1}{2}}g \in L_2(\Omega)$. *Define*

$$(8.21) \quad u^{\infty}(t,x) = |x|^{\frac{1-n}{2}} F(|x|-t, x/|x|), \quad x \in \Omega - \{0\}, \ t \in \mathbb{R}$$

by

$$(8.22) \qquad F(r,\eta) = \mathrm{Re}\{G(r,\eta)\}$$

where G *is defined by* (8.9). *Then*

$$(8.23) \qquad \lim_{t\to+\infty} \|u(t,\cdot) - u^{\infty}(t,\cdot)\|_{L_2(\Omega)} = 0.$$

The proof is the same as that of Theorem 2.8.

COROLLARY 8.4. *The asymptotic wave profile* (8.22) *is characterized by*

$$(8.24) \quad \begin{cases} \hat{F}(\rho,\eta) = \frac{1}{2}(-i\rho)^{\frac{n-1}{2}} \begin{cases} \hat{h}_-(\rho\eta), & \rho > 0 \\ \overline{\hat{h}_-(-\rho\eta)}, & \rho < 0 \end{cases} \\ \qquad\quad = \frac{1}{2}(-i\rho)^{\frac{n-1}{2}} \begin{cases} \hat{f}_-(\rho\eta) + i\,\dfrac{\hat{g}_-(\rho\eta)}{\rho}, & \rho > 0, \\ \hat{f}_+(\rho\eta) + i\,\dfrac{\hat{g}_+(\rho\eta)}{\rho}, & \rho < 0. \end{cases} \end{cases}$$

PROOF. The first equation in (8.24) follows immediately from (8.9) and (8.22). To prove the second equation, note that the distorted plane waves have the property that

$$(8.25) \qquad \overline{w_{\pm}(x,-p)} = w_{\mp}(x,p) \quad \text{for all} \quad x \in \Omega, \quad p \in \mathbb{R}^n.$$

The simplest way to verify this is to note that $\overline{w_{\mp}(x,-p)}$ satisfies conditions (6.9), (6.10), (6.11) which define $w_{\pm}(x,p)$. Thus (8.25) follows from the uniqueness of $w_{\pm}(x,p)$. (8.25) implies that if $h(x)$ and $k(x)$ are in $L_2(\Omega)$ then

$$(8.26) \qquad h(x) = \overline{k(x)} \text{ in } L_2(\Omega) \iff \hat{h}_{+}(-p) = \overline{\hat{k}_{-}(p)} \text{ in } L_2(\mathbb{R}^n).$$

In particular,

$$(8.27) \qquad f(x) = \overline{f(x)} \text{ in } L_2(\Omega) \iff \hat{f}_{+}(-p) = \overline{\hat{f}_{-}(p)} \text{ in } L_2(\mathbb{R}^n).$$

It follows that

$$(8.28) \quad \begin{cases} \overline{\hat{h}_{-}(-\rho\eta)} = \overline{\hat{f}_{-}(-\rho\eta) + i \dfrac{\hat{g}_{-}(-\rho\eta)}{|\rho|}} = \hat{f}_{+}(\rho\eta) - i \dfrac{\hat{g}_{+}(\rho\eta)}{|\rho|} \\[2ex] \qquad = \hat{f}_{+}(\rho\eta) + i \dfrac{\hat{g}_{+}(\rho\eta)}{\rho} \quad \text{for} \quad \rho < 0, \end{cases}$$

which completes the proof of (8.24).

Asymptotic wave functions for solutions wFE in Ω will be derived next. Consider first the complex-valued wave function (8.1) with $h \in D(A^{\frac{1}{2}})$. In this case Theorem 7.5 implies

$$(8.29) \qquad \lim_{t\to+\infty} \| D_k v(t,\cdot) - D_k v_0^{+}(t,\cdot) \|_{L_2(\Omega)} = 0$$

for $k = 0,1,2,\ldots,n$. The cut-off function $j(x)$ can be omitted in (7.52) by the argument given after (8.4) above. Recall that, by (7.56),

$$(8.30)\qquad D_0 v_0^+(t,\cdot) = -i e^{-itA_0^{\frac{1}{2}}} W_+(A^{\frac{1}{2}}h).$$

In particular, $D_0 v_0^+(t,x)$ is a solution in $L_2(\mathbb{R}^n)$ when $h \in D(A^{\frac{1}{2}})$. Moreover, by (8.3)

$$(8.31)\quad (W_+ A^{\frac{1}{2}}h)^\wedge(p) = (\Phi_- A^{\frac{1}{2}}h)(p) = |p|\Phi_- h(p) = |p|\hat{h}_-(p).$$

Thus the profile G_0 of the asymptotic wave function for (8.30) is characterized by

$$(8.32)\ \hat{G}_0(\rho,\eta) = (-i\rho)^{\frac{n-1}{2}} H(\rho)(-i\rho)\hat{h}_-(\rho\eta),\ \rho \in \mathbb{R},\ \eta \in S^{n-1}.$$

The corresponding asymptotic wave function will be denoted by

$$(8.33)\ v_0^\infty(t,x) = |x|^{\frac{1-n}{2}} G_0(|x|-t, x/|x|),\ x \in \Omega - \{0\},\ t \in \mathbb{R},$$

and (8.29) and Theorem 2.6 imply

$$(8.34)\qquad \lim_{t\to+\infty} \| D_0 v(t,\cdot) - v_0^\infty(t,\cdot) \|_{L_2(\Omega)} = 0.$$

Similarly, as in the proof of Theorem 2.10, the derivatives

$$(8.35)\quad D_k v_0^+(t,\cdot) = e^{-itA_0^{\frac{1}{2}}} D_k W_+ h,\quad k = 1,2,\ldots,n$$

are solutions in $L_2(\mathbb{R}^n)$ and

$$(8.36)\qquad (D_k W_+ h)^\wedge(p) = ip_k (W_+ h)^\wedge(p) = ip_k h_-(p).$$

Thus the profiles G_k of the asymptotic wave functions for (8.35) are characterized by

$$(8.37)\quad \hat{G}_k(\rho,\eta) = (-i\rho)^{\frac{n-1}{2}} H(\rho)(i\rho\eta_k)\hat{h}_-(p),\ p \in \mathbb{R},\ \eta \in S^{n-1}$$

or

$$(8.38)\quad \hat{G}_k(\rho,\eta) = -\hat{G}_0(\rho,\eta)\eta_k,\quad k = 1,2,\dots,n.$$

The corresponding asymptotic wave functions will be denoted by

$$(8.39)\quad v_k^\infty(t,x) = |x|^{\frac{1-n}{2}} G_k(|x|-t, x/|x|),\ x \in \Omega - \{0\},\ t \in \mathbb{R},$$

and (8.29) and Theorem 2.6 imply

$$(8.40)\quad \lim_{t\to+\infty} \|D_k v(t,\cdot) - v_k^\infty(t,\cdot)\|_{L_2(\Omega)} = 0,\quad k = 1,2,\dots,n.$$

Collecting these results gives

THEOREM 8.5. *Let* Ω *be an exterior domain such that* $\Omega \in LC$. *Then for all* $h \in D(A^{\frac{1}{2}})$

$$(8.41)\quad \lim_{t\to+\infty} \|D_k v(t,\cdot) - v_k^\infty(t,\cdot)\|_{L_2(\Omega)} = 0,\quad k = 1,2,\dots,n.$$

The corresponding results for solutions wFE of the initial-boundary value problem are given by

COROLLARY 8.6. *Let* Ω *be an exterior domain such that* $\Omega \in LC$. *Let* f *and* g *be real-valued functions such that* $f \in D(A^{\frac{1}{2}}) = L_2^1(\Omega)$ *and* $g \in L_2(\Omega)$. *Define*

$$(8.42)\quad \begin{cases} u_k^\infty(t,x) = |x|^{\frac{1-n}{2}} F_k(|x|-t, x/|x|), \\ x \in \Omega - \{0\},\quad t \in \mathbb{R},\quad k = 0,1,\dots,n \end{cases}$$

by

$$F_k(r,\eta) = \mathrm{Re}\{G_k(r,\eta)\} \tag{8.43}$$

where G_k , $k = 0,1,\ldots,n$ *are defined by* (8.32) *and* (8.38). *Then*

$$\lim_{t\to+\infty} \|D_k u(t,\cdot) - u_k^\infty(t,\cdot)\|_{L_2(\Omega)} = 0, \quad k = 0,1,\ldots,n. \tag{8.44}$$

Each solution wFE in Ω has a finite quantity of energy which is independent of t. The remaining theorems of the lecture describe the asymptotic distribution of this energy, for $t \to \infty$, in subsets of Ω. The results are based on Corollary 8.6 and the form of the functions $u_k^\infty(t,x)$.

In the remainder of the lecture it is assumed without further mention that Ω is an exterior domain such that $\Omega \in LC$ and $u(t,x)$ is a solution wFE in Ω. If $K \subset \Omega$ is any measurable set then the energy in K at time t is defined by

$$E(u,K,t) = \sum_{k=0}^{n} \|D_k u(t,\cdot)\|^2_{L_2(K)} . \tag{8.45}$$

The application of Corollary 8.6 to the calculation of asymptotic energy distributions are based on the following

LEMMA 8.7. *Let* $t \to K(t) \subset \Omega$ *define a family of* Ω *measurable subsets of* Ω *for* $t \in \mathbb{R}$. *Then*

$$(8.46)\quad \begin{cases} \|D_k u(t,\cdot)\|_{L_2(K(t))} \\ = \|u_k^\infty(t,\cdot)\|_{L_2(K(t))} + o(1), \quad t \to \infty, \end{cases}$$

for $k = 0,1,2,\dots,n$ *and hence*

$$(8.47)\quad \begin{cases} E(u,K(t),t) \\ = \sum_{k=0}^{n} \|u_k^\infty(t,\cdot)\|_{L_2(K(t))}^2 + o(1), \quad t \to \infty, \end{cases}$$

where $o(1)$ *tends to zero uniformly with respect to* $\{K(t) : t \in \mathbb{R}\}$.

PROOF. The triangle inequality implies that

$$(8.48)\quad \begin{cases} \left| \|D_k u(t,\cdot)\|_{L_2(K(t))} - \|u_k^\infty(t,\cdot)\|_{L_2(K(t))} \right| \\ \le \|D_k u(t,\cdot) - u_k^\infty(t,\cdot)\|_{L_2(K(t))} \\ \le \|D_k u(t,\cdot) - u_k^\infty(t,\cdot)\|_{L_2(\Omega)} \end{cases}$$

and the last term tends to zero when $t \to \infty$ by Corollary 8.6. Moreover, the last term is independent of the family $\{K(t) : t \in \mathbb{R}\}$.

The first results on asymptotic energy distributions concern the asymptotic concentration of energy in expanding spherical regions of the form

$$(8.49)\quad B(t,\theta_1(t),\theta_2(t)) = \{x: t+\theta_1(t) \le |x| \le t+\theta_2(t)\}.$$

Note that if $\mathbb{R}^n - \Omega \subset B_{r_0}$ then $B(t,\theta_1(t),\theta_2(t)) \subset \Omega$ provided that

$$(8.50) \qquad r_0 - t \le \theta_1(t) \le \theta_2(t) \le +\infty .$$

With this assumption Lemma 8.7 implies

LEMMA 8.8. *Let* $\theta_1(t),\theta_2(t)$ *be any functions of* $t \in \mathbb{R}$ *which satisfy* (8.50) *for all* $t \in \mathbb{R}$. *Then*

$$(8.51) \qquad \begin{cases} E(u,B(t,\theta_1(t),\theta_2(t)),t) \\ \quad = 2\displaystyle\int_{\theta_1(t)}^{\theta_2(t)} \|F_0(r,\cdot)\|^2_{L_2(S^{n-1})} \, dr + o(1) \end{cases}$$

where $o(1)$ *is independent of* $\theta_1(t)$ *and* $\theta_2(t)$.

PROOF. Note first that, by (8.42),

$$(8.52) \qquad \begin{cases} \|u_0^\infty(t,\cdot)\|^2_{L_2(B(t,\theta_1,\theta_2))} = \displaystyle\int_{B(t,\theta_1,\theta_2)} |u_0^\infty(t,x)|^2 dx \\ \quad = \displaystyle\int_{t+\theta_1}^{t+\theta_2} dr \int_{S^{n-1}} |F_0(r-t,\eta)|^2 d\eta \\ \quad = \displaystyle\int_{\theta_1}^{\theta_2} dr \int_{S^{n-1}} |F_0(r,\eta)|^2 d\eta . \end{cases}$$

Next, note that (8.38), (8.43) imply

$$(8.53) \qquad F_k(r,\eta) = -F_0(r,\eta)\eta_k \quad \text{for} \quad k = 1,2,\ldots,n.$$

$$\sum_{k=1}^{n} \|u_k^\infty(t,\cdot)\|^2_{L_2(B(t,\theta_1,\theta_2))}$$

$$= \sum_{k=1}^{n} \int_{B(t,\theta_1,\theta_2)} |u_k^\infty(t,x)|^2 dx$$

(8.54) $$= \sum_{k=1}^{n} \int_{t+\theta_1}^{t+\theta_2} dr \int_{S^{n-1}} |F_k(r-t,\eta)|^2 d\eta$$

$$= \int_{t+\theta_1}^{t+\theta_2} dr \int_{S^{n-1}} |F_0(r-t,\eta)|^2 d\eta$$

$$= \int_{\theta_1}^{\theta_2} dr \int_{S^{n-1}} |F_0(r,\eta)|^2 d\eta$$

because $\eta_1^2 + \eta_2^2 + \ldots + \eta_n^2 = 1$. Combining (8.52), (8.54) and Lemma 8.7 gives (8.51).

THEOREM 8.9. *Let* $\theta_1(t)$, $\theta_2(t)$ *satisfy* (8.50) *for all* $t \in \mathbb{R}$ *and also*

(8.55) $$\lim_{t\to\infty} \theta_1(t) = -\infty, \quad \lim_{t\to\infty} \theta_2(t) = +\infty.$$

Then

(8.56) $$\lim_{t\to+\infty} E(u,B(t,\theta_1(t),\theta_2(t)),t) = E(u,\Omega,0)$$

and hence

(8.57) $$\lim_{t\to+\infty} E(u,\Omega - B(t,\theta_1(t),\theta_2(t)),t) = 0.$$

The proof is based on Lemma 8.8 and

LEMMA 8.10. *For all* $f \in D(A^{\frac{1}{2}}) = L_2^1(\Omega)$ *and* $g \in L_2(\Omega)$

$$E(u,\Omega,0) = 2\,\|F_0\|^2_{L_2(\mathbb{R}\times S^{n-1})}\,. \tag{8.58}$$

PROOF OF THEOREM 8.9. Lemma 8.8, (8.55) and Lemma 8.10 imply

$$\begin{cases} \lim\limits_{t\to+\infty} E(u,B(t,\theta_1(t),\theta_2(t)),t) \\ \\ = 2\,\|F_0\|^2_{L_2(\mathbb{R}\times S^{n-1})} = E(u,\Omega,0)\,. \end{cases} \tag{8.59}$$

Equation (8.57) follows from (8.56) because

$$\begin{cases} E(u,\Omega,0) = E(u,\Omega,t) = E(u,B(t,\theta_1(t),\theta_2(t)),t) \\ \\ \quad + E(u,\Omega-B(t,\theta_1(t),\theta_2(t)),t)\,. \end{cases} \tag{8.60}$$

PROOF OF LEMMA 8.10. The definition of F_0, equations (8.43) and (8.32), implies that

$$\hat{F}_0(\rho,\eta) = \tfrac{1}{2}(-i\rho)^{\frac{n+1}{2}} \begin{cases} \hat{h}_-(\rho\eta), & \rho > 0 \\ \\ \overline{\hat{h}_-(-\rho\eta)}, & \rho < 0. \end{cases} \tag{8.61}$$

Moreover, Parseval's formula and Fubini's theorem imply

$$(8.62)\quad \begin{cases} \|F_0\|^2_{L_2(\mathbb{R}\times S^{n-1})} = \|\hat{F}_0\|^2_{L_2(\mathbb{R}\times S^{n-1})} \\ = \displaystyle\int_{-\infty}^{\infty}\int_{S^{n-1}} |\hat{F}_0(\rho,\eta)|^2 d\eta d\rho . \end{cases}$$

Now (8.61) implies that

$$(8.63)\quad \begin{cases} \displaystyle\int_0^{\infty}\int_{S^{n-1}} |\hat{F}_0(\rho,\eta)|^2 d\eta d\rho = \tfrac{1}{4}\int_0^{\infty}\int_{S^{n-1}} \rho^{n+1}|\hat{h}_-(\rho\eta)|^2 d\eta d\rho \\ = \tfrac{1}{4}\displaystyle\int_{\mathbb{R}^n} |p|^2|\hat{h}_-(p)|^2 dp . \end{cases}$$

Similarly

$$\int_{-\infty}^{0}\int_{S^{n-1}} |\hat{F}_0(\rho,\eta)|^2 d\eta d\rho$$

$$(8.64)\quad \begin{aligned} &= \tfrac{1}{4}\int_{-\infty}^{0}\int_{S^{n-1}} |\rho|^{n+1}|\hat{h}_-(-\rho\eta)|^2 d\eta d\rho \\ &= \tfrac{1}{4}\int_{0}^{\infty}\int_{S^{n-1}} \rho^{n+1}|\hat{h}_-(\rho\eta)|^2 d\eta d\rho \\ &= \tfrac{1}{4}\int_{\mathbb{R}^n} |p|^2|\hat{h}_-(p)|^2 dp . \end{aligned}$$

Adding (8.63) and (8.64) and using (8.62) gives

$$(8.65)\quad \begin{cases} 2\,\|F_0\|^2_{L_2(\mathbb{R}\times S^{n-1})} = \int_{\mathbb{R}^n} |p|^2 |\hat{h}_-(p)|^2 dp \\ \qquad = \int_{\mathbb{R}^n} \big|\, |p|\hat{f}_-(p) + i\hat{g}_-(p)\big|^2 dp. \end{cases}$$

Making the change of variable $p \to -p$ in this integral and using (8.27) gives the alternative representation

$$(8.66)\quad \begin{cases} 2\,\|F_0\|^2_{L_2(\mathbb{R}\times S^{n-1})} = \int_{\mathbb{R}^n} \big|\, |p|\hat{f}_-(-p) + i\hat{g}_-(-p)\big|^2 dp \\ \qquad = \int_{\mathbb{R}^n} \big|\, |p|\overline{\hat{f}_+(p)} + i\overline{g_+(p)}\big|^2 dp \\ \qquad = \int_{\mathbb{R}^n} \big|\, |p|\hat{f}_+(p) - i\hat{g}_+(p)\big|^2 dp. \end{cases}$$

Applying Theorems 6.14 and 6.17 to these two representations gives

$$(8.67)\quad \begin{cases} 2\,\|F_0\|^2_{L_2(\mathbb{R}\times S^{n-1})} = \|A^{\frac{1}{2}} f + i g\|^2_{L_2(\Omega)} \\ \qquad = \|A^{\frac{1}{2}} f - i g\|^2_{L_2(\Omega)}. \end{cases}$$

Finally, adding the last two expressions and using the parallelogram law in $L_2(\Omega)$ gives

$$(8.68)\quad \begin{cases} 2\,\|F_0\|^2_{L_2(\mathbb{R}\times S^{n-1})} = \|A^{\frac{1}{2}} f\|^2_{L_2(\Omega)} + \|g\|^2_{L_2(\Omega)} \\ = \|\nabla f\|^2_{L_2(\Omega)} + \|g\|^2_{L_2(\Omega)} = E(u,\Omega,0)\,. \end{cases}$$

Returning to Theorem 8.9, note that there is no restriction on the rate at which $\theta_1(t)$ and $\theta_2(t)$ tend to infinity. In particular, (8.57) implies the transiency of energy in bounded sets $K \subset \Omega$.

COROLLARY 8.11. *If* K *is any bounded measurable subset of* Ω *then*

$$\lim_{t\to\infty} E(u,K,t) = 0. \tag{8.69}$$

PROOF. Since $K \subset \Omega$ is bounded there is an $r \geq r_0$ such that $K \subset \Omega_r$. In Theorem 8.9 take $\theta_1(t) = r - t \geq r_0 - t$ and $\theta_2(t) = +\infty$. Then $K \subset \Omega_r \subset \Omega - B(t,\theta_1(t),\theta_2(t))$ for all t and hence

$$0 \leq E(u,K,t) \leq E(u,\Omega - B(t,\theta_1(t),\theta_2(t)),t) \tag{8.70}$$

for all t. Making $t \to \infty$ and using (8.57) gives (8.69).

Lemma 8.8 also implies that solutions wFE are, to any preassigned degree of precision, concentrated in an expanding spherical zone of constant thickness for all t large enough. The thickness of the zone is determined by the initial values f,g. The result may be stated as follows.

COROLLARY 8.12. *Let* $f \in L_2^1(\Omega)$, $g \in L_2(\Omega)$ *and* $\epsilon > 0$ *be given. Then there exist constants* $\theta_1 = \theta_1(f,g,\epsilon)$, $\theta_2 = \theta_2(f,g,\epsilon)$ *and* $t_0 = t_0(f,g,\epsilon)$ *such that*

$$E(u,\Omega,0) - \epsilon \leq E(u,B(t,\theta_1,\theta_2),t) \leq E(u,\Omega,0) \tag{8.71}$$

for all $t \geq t_0$.

PROOF. Note that (8.71) is equivalent to

$$(8.72)\quad \begin{cases} |E(u,\Omega,0) - E(u,B(t,\theta_1,\theta_2),t)| \le \epsilon \\ \text{for all} \quad t \ge t_0. \end{cases}$$

Now by Lemma 8.10,

$$(8.73)\quad \begin{cases} |E(u,\Omega,0) - E(u,B(t,\theta_1,\theta_2),t)| \\ = \left|2\int_{-\infty}^{\infty} \|F_0(r,\cdot)\|^2 dr - E(u,B(t,\theta_1,\theta_2),t)\right| \\ \le 2\int_{-\infty}^{\theta_1} \|F_0(r,\cdot)\|^2 dr + 2\int_{\theta_2}^{\infty} \|F_0(r,\cdot)\|^2 dr \\ + \left|2\int_{\theta_1}^{\theta_2} \|F_0(r,\cdot)\|^2 dr - E(u,B(t,\theta_1,\theta_2),t)\right| \end{cases}$$

for all t. Lemma 8.8 implies that there is a $t_0 = t_0(f,g,\epsilon)$, independent of θ_1, θ_2, such that the last term in (8.73) is less than $\epsilon/2$ for all $t \ge t_0$. Moreover, there exist constants $\theta_1 = \theta_1(f,g,\epsilon)$, $\theta_2 = \theta_2(f,g,\epsilon)$ such that the sum of the first and second integrals on the right-hand side of (8.73) is less than $\epsilon/2$. With these choices (8.73) implies (8.72) and hence (8.71).

Lemma 8.7 implies still more precise results concerning the asymptotic concentration of solutions wFE. The nature of the sets in which $u(t,x)$ is concentrated for large t is determined by the initial state f,g through the function F_0. To formulate this idea let $\epsilon > 0$ be given and let $S = S(f,g,\epsilon) \subset \mathbb{R} \times S^{n-1}$ be any set such that

$$(8.74)\quad \begin{cases} \|F_0\|^2_{L_2(\mathbb{R}\times S^{n-1})} - \frac{\epsilon}{4} \le \int_S |F_0(r,\eta)|^2 dr d\eta \\ \qquad\qquad \le \|F_0\|^2_{L_2(\mathbb{R}\times S^{n-1})} . \end{cases}$$

Introduce the notation

$$(8.75)\qquad S_t = \{(r+t,\eta) : (r,\eta) \in S\} \subset \mathbb{R} \times S^{n-1},$$

$$(8.76)\qquad S^{\pm} = S \cap \{(r,\eta) : \pm r \ge 0\}$$

and

$$(8.77)\qquad K_t = \{x = r\eta : (r,\eta) \in S_t^{+}\} \subset \mathbb{R}^n .$$

Then the following refinement of Corollary 8.12 holds.

COROLLARY 8.13. *Let* $f \in L_2^1(\Omega)$, $g \in L_2(\Omega)$ *and* $\epsilon > 0$ *be given. Let* $S = S(f,g,\epsilon)$ *be any subset of* $\mathbb{R} \times S^{n-1}$ *such that* (8.74) *holds. Then there exists a* $t_0 = t_0(f,g,\epsilon)$ *such that*

$$(8.78)\qquad E(u,\Omega,0) - \epsilon \le E(u,\Omega\cap K_t , t) \le E(u,\Omega,0)$$

for all $t \ge t_0$.

PROOF. (8.78) is equivalent to

$$(8.79)\qquad |E(u,\Omega,0) - E(u,\Omega\cap K_t , t)| \le \epsilon \quad \text{for all} \quad t \ge t_0 ;$$

by Lemma 8.10

$$(8.80)\qquad \begin{cases} |E(u,\Omega,0) - E(u,\Omega\cap K_t\,,\,t)| \\ \le \left|2\,\|F_0\|^2 - 2\displaystyle\int_S |F_0(r,\eta)|^2 dr d\eta\right| \\ + \left|2\displaystyle\int_S |F_0(r,\eta)|^2 dr d\eta - E(u,\Omega\cap K_t\,,\,t)\right|. \end{cases}$$

The first term on the right is less than $\epsilon/2$ by the choice (8.74) of S. To estimate the second term note that, by Lemma 8.7,

$$(8.81)\qquad E(u,\Omega\cap K_t\,,\,t) = \sum_{k=0}^{n} \|u_k^\infty(t,\cdot)\|^2_{L_2(\Omega\cap K_t)} + o(1).$$

The asymptotic wave functions $u_k^\infty(t,\cdot)$ have an extension to $L_2(\mathbb{R}^n)$ and tend to zero in $L_2(K)$ for any bounded $K \subset \mathbb{R}^n$. Thus (8.81) implies

$$E(u,\Omega\cap K_t\,,\,t) = \sum_{k=0}^{n} \|u_k^\infty(t,\cdot)\|^2_{L_2(K_t)} + o(1)$$

$$= \sum_{k=0}^{n} \int_{K_t} |F_k(|x|-t,x/|x|)|^2 |x|^{1-n} dx + o(1)$$

$$(8.82)\qquad = 2\int_{K_t} |F_0(|x|-t,x/|x|)|^2 |x|^{1-n} dx + o(1)$$

$$= 2\int_{S_t^+} |F_0(r-t,\eta)|^2 dr d\eta + o(1)$$

$$= 2\int_{S\cap\{(r,\eta)\,:\, r\ge -t\}} |F_0(r,\eta)|^2 dr d\eta + o(1).$$

Note that

$$(8.83)\quad 0 \le \int_{S\cap\{(r,\eta):r\le -t\}} |F_0(r,\eta)|^2 dr d\eta \le \int_{-\infty}^{-t}\int_{S^{n-1}} |F_0(r,\eta)|^2 d\eta dr$$

and the last integral tends to zero when $t \to \infty$. Thus (8.82) and (8.83) imply that

$$(8.84)\qquad E(u,\Omega\cap K_t,t) = 2\int_S |F_0(r,\eta)|^2 dr d\eta + o(1),\quad t\to\infty.$$

Thus there is a $t_0 = t_0(f,g,\epsilon)$ such that the last term in (8.80) is less than $\epsilon/2$ for all $t \ge t_0$. This completes the proof.

Now consider a cone in $\mathbb{R}^n$ with vertex at the origin; that is, a set of the form

$$(8.85)\qquad C = \{x = r\eta : r > 0 \text{ and } \eta \in C_0\}$$

where C_0 is a Lebesgue-measurable subset of S^{n-1}. The asymptotic distribution of energy in cones is described by

THEOREM 8.14. *The asymptotic energy distribution*

$$(8.86)\qquad E^\infty(u,\Omega\cap C) = \lim_{t\to\infty} E(u,\Omega\cap C,t)$$

exists for each cone C *and is given by*

$$(8.87)\qquad E^\infty(u,\Omega\cap C) = \int_C \big||p|\hat f_-(p) + i\hat g_-(p)\big|^2 dp.$$

A first step in the proof of Theorem 8.14 is

LEMMA 8.15. *For each cone* C

$$(8.88)\qquad E^\infty(u,\Omega\cap C) = 2\,\|F_0\|^2_{L_2(\mathbb{R}\times C_0)}\,.$$

PROOF OF LEMMA 8.15. Let

$$(8.89)\quad C_h' = C\cap B_h' = \{x = r\eta\colon r \ge h \text{ and } \eta\in C_0\}.$$

Then if $h \ge r_0$, $\Omega\cap C = (\Omega_h\cap C)\cup C_h'$ since $\mathbb{R}^n - \Omega\subset B_{r_0}$ and hence

$$(8.90)\qquad E(u,\Omega\cap C,t) = E(u,\Omega_h\cap C,t) + E(u,C_h',t)\,.$$

Now $E(u,\Omega_h\cap C,t) = o(1)$ for $t\to\infty$ by Corollary 8.11. Thus combining (8.90) and Lemma 8.7 gives

$$(8.91)\qquad E(u,\Omega\cap C,t) = \sum_{k=0}^{n}\|u_k^\infty(t,\cdot)\|^2_{L_2(C_h')} + o(1)\,.$$

Now

$$(8.92)\quad \begin{cases} \|u_0^\infty(t,\cdot)\|^2_{L_2(C_h')} = \displaystyle\int_{C_h'} |F_0(|x|-t,x/|x|)|^2|x|^{1-n}dx \\ = \displaystyle\int_h^\infty\int_{C_0} |F_0(r-t,\eta)|^2 d\eta dr = \int_{h-t}^\infty\int_{C_0} |F_0(r,\eta)|^2 d\eta dr \\ = \|F_0\|^2_{L_2(\mathbb{R}\times C_0)} + o(1),\quad t\to+\infty. \end{cases}$$

Similarly

$$(8.93)\quad \begin{cases} \displaystyle\sum_{k=1}^{n}\|u_k^\infty(t,\cdot)\|^2_{L_2(C_h')} = \sum_{k=1}^{n}\int_h^\infty\int_{C_0} |F_k(r-t,\eta)|^2 d\eta dr \\ = \displaystyle\int_h^\infty\int_{C_0} |F_0(r-t,\eta)|^2 d\eta dr = \|F_0\|^2_{L_2(\mathbb{R}\times C_0)} + o(1)\,. \end{cases}$$

Adding (8.92) and (8.93) and using (8.91) gives

$$(8.94)\qquad E(u,\Omega\cap C,t) = 2\,\|F_0\|^2_{L_2(\mathbb{R}\times C_0)} + o(1), \quad t\to +\infty,$$

which is equivalent to (8.88).

PROOF OF THEOREM 8.14. It will be shown that

$$(8.95)\qquad 2\,\|F_0\|^2_{L_2(\mathbb{R}\times C_0)} = \int_C \big|\,|p|\hat{f}_-(p) + i\hat{g}_-(p)\big|^2 dp.$$

To this end note that $F_0 = \frac{1}{2}(G_0 + \overline{G}_0)$ where G_0 and $\overline{G}_0$ are orthogonal in $L_2(\mathbb{R}\times C_0)$. This follows from Parseval's formula in $L_2(\mathbb{R}, L_2(C_0))$ and the fact that $\hat{G}_0(\rho,\eta)$ and $\overline{\hat{G}_0(-\rho,\eta)}$, the transform of $\overline{G}_0$, have disjoint supports. Thus

$$(8.96)\qquad \begin{cases} 2\,\|F_0\|^2_{L_2(\mathbb{R}\times C_0)} = \frac{1}{2}\|G_0\|^2_{L_2(\mathbb{R}\times C_0)} \\ \quad + \frac{1}{2}\|\overline{G}_0\|_{L_2(\mathbb{R}\times C_0)} = \|G_0\|^2_{L_2(\mathbb{R}\times C_0)} \\ \quad = \|\hat{G}_0\|^2_{L_2(\mathbb{R}\times C_0)} = \displaystyle\int_0^\infty\int_{C_0} |\hat{G}_0(\rho,\eta)|^2 d\eta\, d\rho \\ \quad = \displaystyle\int_0^\infty\int_{C_0} \rho^{n+1}|\hat{h}_-(\rho\eta)|^2 d\eta\, d\rho = \int_C |p|^2|\hat{h}_-(p)|^2 dp. \end{cases}$$

This equivalent to (8.95) because

$$(8.97)\qquad \begin{cases} |p|\hat{h}_-(p) = |p|(\Phi_- h)(p) = \Phi_-(A^{\frac{1}{2}}h) = \Phi_-(A^{\frac{1}{2}}f + ig) \\ \qquad = \Phi_-(A^{\frac{1}{2}}f) + i\Phi_- g = |p|\hat{f}_-(p) + i\hat{g}_-(p). \end{cases}$$

COROLLARY 8.16. *The limiting distribution* (8.87)

can be written

$$(8.98)\quad \begin{cases} E^{\infty}(u,\Omega\cap C) = \displaystyle\int_C |\Phi_-(A^{\frac{1}{2}}f + ig)|^2 dp \\ \qquad\qquad\quad\; = \displaystyle\int_C |\Phi W_+(A^{\frac{1}{2}}f + ig)|^2 dp. \end{cases}$$

This is immediate from (8.87), (8.97) and the relation $W_+ = \Phi^*\Phi_-$.

Every cone in $\mathbb{R}^n$ can be written in the form $C + \bar{x}$ where $\bar{x} \in \mathbb{R}^n$ and C is a cone with vertex at the origin. This shift by $\bar{x}$ has no effect on the asymptotic energy distribution. This will be stated as

COROLLARY 8.17. *For each cone* C

$$(8.99)\quad E^{\infty}(u,\Omega\cap(C+\bar{x})) = E^{\infty}(u,\Omega\cap C) \text{ for every } \bar{x} \in \mathbb{R}^n.$$

PROOF. A proof by direct calculation, using Lemma 8.7, is quite tedious. However, note that under a change of coordinates $x \to \tilde{x} = x - \bar{x}$ boundary value problems for Ω become boundary value problems for $\Omega - \bar{x}$. It is not difficult to check that the eigenfunctions for $\Omega - \bar{x}$ in the new coordinate system are $\tilde{w}_{\pm}(\tilde{x},p) = e^{-i\bar{x}\cdot p}w_{\pm}(x,p)$. Moreover, if $\tilde{u}(t,\tilde{x}) = u(t,x)$ then $\tilde{u}(t,\tilde{x})$ is a solution wFE for $\Omega - \bar{x}$ and

$$(8.100)\quad E(u,\Omega\cap(C+\bar{x}),t) = E(\tilde{u},(\Omega-\bar{x})\cap C,t) \quad \text{for } t \in \mathbb{R}.$$

It follows that

$$(8.101)\quad \begin{cases} E^{\infty}(u,\Omega\cap(C+\bar{x})) = E^{\infty}(\tilde{u},(\Omega-\bar{x})\cap C) \\ \qquad = \displaystyle\int_C \left|\,|p|\hat{\tilde{f}}_-(p) + i\hat{\tilde{g}}_-(p)\right|^2 dp. \end{cases}$$

But

$$(8.102)\quad \begin{cases} \hat{\tilde{f}}_-(p) = \displaystyle\int_{\Omega-\bar{x}} \overline{\tilde{w}_-(\tilde{x},p)}\,\tilde{f}(\tilde{x})\,d\tilde{x} \\ \qquad = \displaystyle\int_{\Omega} \overline{e^{-ip\cdot\bar{x}}w_-(x,p)}\,f(x)\,dx = e^{ip\cdot\bar{x}}\hat{f}_-(p) \end{cases}$$

and similarly $\hat{\tilde{g}}_-(p) = e^{ip\cdot\bar{x}}\hat{g}_-(p)$. Thus (8.101) implies (8.99).

Now consider a slab in $\mathbb{R}^n$; that is, a set of the type

$$(8.103)\qquad S = \{x\colon d_1 \le x\cdot\bar{x} \le d_2\}$$

where d_1 and $d_2 \ge d_1$ are constants and $\bar{x}$ is a unit vector. S can be written as the difference of two half-spaces. Thus if

$$(8.104)\quad \begin{cases} H = \{x\colon x\cdot\bar{x} \ge 0\} \\ H_1 = H + d_1\bar{x} \\ H_2 = H + d_2\bar{x} \end{cases}$$

then

$$(8.105)\qquad S = H_1 - H_2.$$

It follows that

$$(8.106)\qquad E(u,\Omega\cap H_1,t) = E(u,\Omega\cap H_2,t) + E(u,\Omega\cap S,t)$$

and hence

$$E^{\infty}(u,\Omega\cap H_1) = E^{\infty}(u,\Omega\cap H_2) + E^{\infty}(u,\Omega\cap S). \tag{8.107}$$

But Corollary 8.17 implies that

$$E^{\infty}(u,\Omega\cap H_1) = E^{\infty}(u,\Omega\cap H_2) = E^{\infty}(u,\Omega\cap H). \tag{8.108}$$

Combining this with (8.107) proves

COROLLARY 8.18. *For every slab* S

$$E^{\infty}(u,S) = 0. \tag{8.109}$$

APPENDIX. AN EXISTENCE PROOF FOR THE WAVE OPERATORS BASED ON THE ABSTRACT THEORY OF SCATTERING OF M. S. BIRMAN

The purpose of this appendix is to describe an abstract operator-theoretic existence theorem for wave operators due to M. S. Birman and its application to the proof of Theorem 5.6. The following notation is used to formulate the abstract theorem.

$\mathcal{H}_0$, $\mathcal{H}$	denote separable Hilbert spaces.
H_0 , H	denote selfadjoint operators on $\mathcal{H}_0$ and $\mathcal{H}$, respectively.
$\Pi_0(\lambda)$, $\Pi(\lambda)$	denote the spectral families of H_0 and H, respectively.
$\mathcal{H}_0^{ac}$, $\mathcal{H}^{ac}$	denote the subspaces of absolute continuity of H_0 and H, respectively.
P_0^{ac}, P^{ac}	denote the orthogonal projections of $\mathcal{H}_0$ and $\mathcal{H}$ onto $\mathcal{H}_0^{ac}$ and $\mathcal{H}^{ac}$, respectively.

$$\Pi_0^{ac}(\lambda) = P_0^{ac}\Pi_0(\lambda), \quad \Pi^{ac}(\lambda) = P^{ac}\Pi(\lambda).$$

The following classes of linear operators are needed:

$B(\mathcal{H}_0 , \mathcal{H})$	the class of bounded linear operators from $\mathcal{H}_0$ to $\mathcal{H}$,

$B_0(H_0, H)$	the class of compact linear operators from H_0 to H,
$B_1(H_0, H)$	the class of nuclear (trace-class) linear operators from H_0 to H.

The classes satisfy the inclusion relations

$$B_1(H_0, H) \subset B_0(H_0, H) \subset B(H_0, H).$$

Let

$J_0 \in B(H_0, H)$,

M denote a closed linear subspace of H_0,

P_M denote the orthogonal projection of M onto H_0.

The wave operators

$$W_\pm(H, H_0, J_0, M) = \underset{t \to \pm\infty}{\text{s-lim}}\, e^{itH} J_0 e^{-itH_0} P_M \tag{A.1}$$

will be associated with the quadruple (H, H_0, J_0, M) whenever the strong limit in (A.1) exists. The definitions of the concepts mentioned above may be found in the book of T. Kato [18].

Abstract existence theorems for wave operators of the form (A.1) have been developed by M. S. Birman and his collaborators. The basic theorems were given in [2,4]. A number of variants, or corollaries, of these theorems have been developed and applied to specific classes of differential operators [3,4,24,25,34]. The hypotheses of each of these theorems are adapted to the particular applications to be studied. Here it will be convenient to apply the following corollary of the Birman theory due to W. C. Lyford [25].

THEOREM A.1. *Assume that* H, H_0 *and* J_0 *satisfy*

(A.2) $$J_0 D(H_0) \subset D(H), \quad J_0^* D(H) \subset D(H_0).$$

Moreover, let $\{I_m\}$ *be a family of disjoint bounded open intervals such that*

(A.3) $$\bigcup_{m=1}^{\infty} I_m = \mathbb{R} - Z,$$

where Z *is a Lebesgue null set, and assume that for* $m = 1,2,3, \ldots,$

(A.4) $$(HJ_0 - J_0H_0)\Pi_0^{ac}(I_m) \in B_1(\mathcal{H}_0, \mathcal{H})$$

(A.5) $$(J_0^*J_0 - 1)\Pi_0^{ac}(I_m) \in B_0(\mathcal{H}_0, \mathcal{H}_0)$$

(A.6) $$(J_0J_0^* - 1)\Pi^{ac}(I_m) \in B_0(\mathcal{H}, \mathcal{H}).$$

Then the wave operators

(A.7) $$W_\pm = W_\pm(H_0, H, J_0^*, \mathcal{H}^{ac})$$

and

(A.8) $$W_\pm^0 = W_\pm(H, H_0, J_0, \mathcal{H}_0^{ac})$$

exist. Moreover, $W_\pm : \mathcal{H} \to \mathcal{H}_0$ *is partially isometric with initial set* $\mathcal{H}^{ac}$ *and final set* $\mathcal{H}_0^{ac}$. *This means that*

(A.9) $$W_\pm^* W_\pm = P^{ac} \quad \text{and} \quad W_\pm W_\pm^* = P_0^{ac}.$$

Similarly, $W_\pm^0 : \mathcal{H}_0 \to \mathcal{H}$ *is partially isometric with initial set* $\mathcal{H}_0^{ac}$ *and final set* $\mathcal{H}^{ac}$*:*

$$(A.10) \qquad (W_\pm^0)^* W_\pm^0 = P_0^{ac} \quad \text{and} \quad W_\pm^0 (W_\pm^0)^* = P^{ac}.$$

In addition,

$$(A.11) \qquad W_\pm^* = W_\pm^0 .$$

Finally, the invariance principle holds:

$$(A.12) \qquad W_\pm = W_\pm(\phi(H_0),\ \phi(H),\ J_0^*,\ \mathcal{H}^{ac})$$

for all continuous monotone increasing functions $\phi(\lambda)$.

Theorem A.1 is derived from Birman's basic theorem [2, Theorem 4.4] in [25]. The class of partially isometric operators is discussed in [18]. The invariance principle (A.12) also holds for large classes of discontinuous monotone increasing functions; see [18].

Theorem A.1 will be applied to the operators A_0 and A defined in Lectures 2 and 3. The following identifications will be made.

$$(A.13) \qquad \mathcal{H}_0 = L_2(\mathbb{R}^n), \quad \mathcal{H} = L_2(\Omega)$$

$$(A.14) \qquad H_0 = A_0, \quad H = A.$$

Note that in this case, by Theorem 5.3,

$$(A.15) \qquad \mathcal{H}_0^{ac} = L_2(\mathbb{R}^n), \quad \mathcal{H}^{ac} = L_2(\Omega)$$

and hence

(A.16) $$P_0^{ac} = 1,\ P^{ac} = 1$$

and

(A.17) $$\Pi_0^{ac}(\lambda) = \Pi_0(\lambda),\quad \Pi^{ac}(\lambda) = \Pi(\lambda).$$

Finally, J_0 is defined by

(A.18) $$J_0 = J^*$$

where $J: L_2(\Omega) \to L_2(\mathbb{R}^n)$ is the operator defined by (6.33); i.e.,

(A.19) $$Jf(x) = \begin{cases} j(x)f(x), & x \in \Omega, \\ 0 & , \ x \in \mathbb{R}^n - \Omega \end{cases}$$

and hence

(A.20) $$J^*f(x) = j(x)f(x),\quad x \in \Omega.$$

Hypotheses (A.2)-(A.6) will be verified for these operators and any choice of the intervals I_m.

The domains of the operators A_0 and A are $D(A_0) = L_2(\Delta,\mathbb{R}^n)$ and $D(A) = L_2^N(\Delta,\Omega)$. Thus, hypothesis (A.2) becomes

(A.21) $$J^*L_2(\Delta,\mathbb{R}^n) \subset L_2^N(\Delta,\Omega),\quad JL_2^N(\Delta,\Omega) \subset L_2(\Delta,\mathbb{R}^n).$$

The validity of these conditions is evident from (A.19) and (A.20) because $j(x)$ vanishes in a neighborhood of $\mathbb{R}^n - \Omega$, and $L_2(\Delta,\mathbb{R}^n) = L_2^2(\mathbb{R}^n)$.

To verify (A.4), (A.5) and (A.6) it is necessary to calculate the operators $AJ^* - J^*A_0$, JJ^* and J^*J. Note first that $A = -\Delta$ on $D(A)$. Hence, for $u \in D(A_0)$

$$(A.22)\quad \begin{cases} AJ^*u(x) = -\Delta\{j(x)u(x)\} \\ \qquad = -\{j(x)\Delta u(x)+2\nabla j(x)\cdot\nabla u(x)+\Delta j(x)u(x)\} \end{cases}$$

and

$$(A.23)\qquad J^*A_0u(x) = j(x)\{-\Delta u(x)\} = -j(x)\Delta u(x).$$

Thus

$$(A.24)\qquad (AJ^* - J^*A_0)u(x) = -2\nabla j(x)\cdot\nabla u(x) - \Delta j(x)\cdot u(x)$$

for all $u \in D(A_0)$. The spectral family $\Pi_0(\lambda)$ was defined by (2.24). It follows that

$$(A.25)\qquad \Pi_0(I)u(x) = \frac{1}{(2\pi)^{n/2}}\int_{|p|^2\in I} e^{ix\cdot p}\hat{u}(p)\,dp$$

where I is any interval. Combining (A.24) and (A.25) gives

$$(A.26)\quad \begin{cases} (AJ^* - J^*A_0)\Pi_0(I)u(x) \\ = \dfrac{1}{(2\pi)^{n/2}}\displaystyle\int_{|p|^2\in I}\{-2ip\cdot\nabla j(x) - \Delta j(x)\}e^{ix\cdot p}\hat{u}(p)\,dp. \end{cases}$$

Finally, note that combining (A.19) and (A.20) gives

$$(A.27)\qquad JJ^*u(x) = j^2(x)u(x)\quad \text{for all}\quad x \in \mathbb{R}^n$$

and

$$(A.28)\qquad J^*Ju(x) = j^2(x)u(x)\quad \text{for all}\quad x \in \Omega.$$

Thus

$$
\text{(A.29)}\qquad \begin{cases} (JJ^* - 1)\Pi_0(I)u(x) \\ = (j^2(x) - 1)\Pi_0(I)u(x) \quad \text{for} \quad x \in \mathbb{R}^n \end{cases}
$$

and

$$
\text{(A.30)}\qquad \begin{cases} (J^*J - 1)\Pi(I)u(x) \\ = (j^2(x) - 1)\Pi(I)u(x) \quad \text{for} \quad x \in \Omega . \end{cases}
$$

Note that (A.26) implies that $(AJ^* - J^*A_0)\Pi_0(I) = T\Phi$ where Φ is the Fourier transform and T is an integral operator from $L_2(\mathbb{R}^n)$ to $L_2(\Omega)$ of the form

$$
\text{(A.31)}\qquad Tf(x) = \int_K \tau(x,p) f(p)dp, \quad x \in \Omega,
$$

where

$$
\text{(A.32)}\qquad K = \{p: |p|^2 \in \bar{I}\} \subset \mathbb{R}^n
$$

is a compact set and

$$
\text{(A.33)}\qquad \begin{cases} \tau(x,p) = \dfrac{1}{(2\pi)^{n/2}} \{-2ip \cdot \nabla j(x) - \Delta j(x)\} e^{ix\cdot p}, \\ x \in \Omega, \quad p \in K. \end{cases}
$$

Criteria for integral operators of the form (A.31) to be nuclear have been given by W. F. Stinespring [37]. One of the results of [37] will be used to verify (A.4). The following notation will be used. H will denote a separable Hilbert space. $L_2(\mathbb{R}^n, H)$ will denote the Hilbert space of functions $u: \mathbb{R}^n \to H$ which are Bochner measurable and Bochner square-integrable on $\mathbb{R}^n$;

see [12] for definitions of these terms. Similarly, let $L_2^m(\mathbb{R}^n,\mathcal{H}) = \{u\colon D^\alpha u \in L_2(\mathbb{R}^n,\mathcal{H})$ for $0 \le |\alpha| \le m\}$. Integral operators $T\colon L_2(\mathbb{R}^n) \to \mathcal{H}$ of the form

$$\text{(A.34)} \qquad Tf = \int_K \tau(p) f(p)\,dp, \quad f \in L_2(\mathbb{R}^n),$$

will be considered, where $K \subset \mathbb{R}^n$ is a measurable set, $\tau \in L_2(\mathbb{R}^n,\mathcal{H})$, and the integral is a Bochner integral. The following theorem concerning operators of the form (A.34) is sufficient for the application to the operator defined by (A.26). It is a special case of [37, Theorem 2].

THEOREM A.2. *Assume that* K *is compact,* supp τ *is compact and* $\tau \in L_2^m(\mathbb{R}^n,\mathcal{H})$ *for some* $m > n/2$. *Then* (A.34) *defines an operator* $T \in B_1(L_2(\mathbb{R}^n),\mathcal{H})$.

To apply Theorem A.2 to the operator T defined by (A.31), (A.32), (A.33) note that the kernel $\tau(x,p)$ of (A.33) can be defined for $p \in \mathbb{R}^n - K$ in any way that is convenient. Define

$$\text{(A.35)} \quad \begin{cases} \tau(x,p) = \dfrac{1}{(2\pi)^{n/2}} \{-2ip \cdot \nabla j(x) - \Delta j(x)\} e^{ix\cdot p} \phi(p), \\ x \in \Omega, \quad p \in \mathbb{R}^n \end{cases}$$

where $\phi \in \mathcal{D}(\mathbb{R}^n)$ and $\phi(p) = 1$ for all $p \in K$. With this definition it is clear that $\tau \in L_2^m(\mathbb{R}^n,L_2(\Omega))$ for every positive integer m because $\tau \in \mathcal{D}(\Omega \times \mathbb{R}^n)$ and supp $\tau(\cdot,p) \subset \Omega_{r_0+1}$ for all $p \in \mathbb{R}^n$. Thus Theorem A.2 implies that

$$(A.36) \qquad (AJ^* - J^*A_0)\Pi_0(I) = T\Phi \in B_1(L_2(\mathbb{R}^n), L_2(\Omega))$$

because $T \in B_1(L_2(\mathbb{R}^n), L_2(\Omega))$ and $\Phi \in B(L_2(\mathbb{R}^n), L_2(\mathbb{R}^n))$. This completes the verification of condition (A.4).

Now consider condition (A.6). To verify it the operator $(JJ^* - 1)\Pi(I)$ defined by (A.30) must be shown to belong to the class $B_0(L_2(\Omega), L_2(\Omega))$. To this end let $\{u_m\}$ be a bounded sequence in $L_2(\Omega)$. It must be shown that

$$(A.37) \qquad (j^2(x) - 1)\Pi(I)u_m(x), \quad m = 1,2,3,\ldots$$

has a subsequence which converges in $L_2(\Omega)$. Now $\operatorname{supp}(j^2(x) - 1) \subset B(r_0 + 1)$. Hence it is enough to show that $\{\Pi(I)u_m\}$ has a subsequence which converges in $L_2(\Omega_{r_0+1})$. Assume that C is a bound for $\{u_m\}$:

$$(A.38) \qquad \|u_m\|_{L_2(\Omega)} \leq C, \quad m = 1,2,3,\ldots .$$

Then the functions $v_m = \Pi(I)u_m \in D(A) = L_2^N(\Delta,\Omega)$ and the spectral theorem implies

$$(A.39) \begin{cases} \|Av_m\|^2_{L_2(\Omega)} = \|A\Pi(I)u_m\|^2_{L_2(\Omega)} = \int_I \lambda^2 d\,\|\Pi(\lambda)u_m\|^2_{L_2(\Omega)} \\ \leq M^2 \|u_m\|^2_{L_2(\Omega)} \leq M^2C^2 \end{cases}$$

where $M = \max_{\lambda\in\bar{I}} |\lambda|$. Moreover, the generalized Neumann condition (3.10), applied to $u = v_m \in L_2^N(\Delta,\Omega)$ and $v = \overline{v_m}$, gives

$$\text{(A.40)}\quad \begin{cases} \|\nabla v_m\|^2_{L_2(\Omega)} = -\int_\Omega \bar{v}_m \Delta v_m dx = (v_m , Av_m)_{L_2(\Omega)} \\ \le \|v_m\|_{L_2(\Omega)} \|Av_m\|_{L_2(\Omega)} \\ \le \tfrac{1}{2}\left(\|v_m\|^2_{L_2(\Omega)} + \|Av_m\|^2_{L_2(\Omega)}\right) \le \tfrac{1}{2}(C^2 + M^2C^2) . \end{cases}$$

Thus $\{v_m\}$ is bounded in $L_2^1(\Omega)$ and hence the local compactness theorem, Theorem 4.3, implies that $\{v_m\} = \{\Pi(I)u_m\}$ has a convergent subsequence in $L_2(\Omega_{r_0+1})$. This completes the verification of condition (A.6).

The verification of condition (A.5) is the same as for (A.6). Thus all the conditions of Theorem A.1 have been verified for $(H,H_0,J_0) = (A,A_0,J^*)$. This proves

THEOREM A.3. *Let* Ω *be an exterior domain such that* $\Omega \in LC$. *Then the wave operators*

$$\text{(A.41)}\qquad W_\pm = W_\pm(A_0 , A,J,L_2(\Omega))$$

and

$$\text{(A.42)}\qquad W_\pm^0 = W_\pm(A,A_0 , J^*,L_2(\mathbb{R}^n))$$

exist. Moreover, $W_\pm: L_2(\Omega) \to L_2(\mathbb{R}^n)$ *and* $W_\pm^0: L_2(\mathbb{R}^n) \to L_2(\Omega)$ *are unitary,*

$$\text{(A.43)}\qquad W_\pm^* = W_\pm^0 ,$$

and the invariance principle holds:

$$\text{(A.44)}\qquad W_\pm = W_\pm(\phi(A_0), \phi(A), J, L_2(\Omega)).$$

COROLLARY A.4. *Theorem* 5.6 *is valid; that is, the wave operators*

$$\text{(A.45)} \qquad W_\pm(A_0^{\frac12}, A^{\frac12}, J_\Omega) = W_\pm(A_0^{\frac12}, A^{\frac12}, J_\Omega, L_2(\Omega))$$

exist and are unitary. In fact

$$\text{(A.46)} \qquad W_\pm(A_0^{\frac12}, A^{\frac12}, J_\Omega, L_2(\Omega)) = W_\pm,$$

where $W_\pm$ *is defined by* (A.41). *Moreover,*

$$\text{(A.47)} \qquad \Pi(\lambda) = W_\pm^* \Pi_0(\lambda) W_\pm \quad \textit{for all} \quad \lambda \in \mathbb{R}.$$

PROOF. Note first (A.44) with $\phi(\lambda) = \lambda^{\frac12}$ for $\lambda > 0$ and $\phi(\lambda) = 0$ for $\lambda \leq 0$ implies that

$$\text{(A.48)} \qquad W_\pm = W_\pm(A_0^{\frac12}, A^{\frac12}, J, L_2(\Omega)).$$

Moreover,

$$\text{(A.49)} \qquad (J_\Omega - J)f(x) = \begin{cases} (1 - j(x))f(x), & x \in \Omega \\ 0 & , \ x \in \mathbb{R}^n - \Omega. \end{cases}$$

In particular, $\operatorname{supp}(J_\Omega - J)f \subset B(r_0 + 1)$. Now

$$\text{(A.50)} \qquad \begin{cases} e^{itA_0^{\frac12}} J_\Omega e^{-itA^{\frac12}} u \\ = e^{itA_0^{\frac12}} J e^{-itA^{\frac12}} u + e^{itA_0^{\frac12}} (J_\Omega - J) e^{-itA^{\frac12}} u \end{cases}$$

and the last term tends to zero in $L_2(\mathbb{R}^n)$ when $t \to \pm\infty$, for any $u \in L_2(\Omega)$, by Theorem 5.5. Thus making $t \to \pm\infty$ in (A.50) gives

$$\text{(A.51)}\quad W_{\pm}(A_0^{\frac{1}{2}}, A^{\frac{1}{2}}, J_{\Omega}, L_2(\Omega)) = W_{\pm}(A_0^{\frac{1}{2}}, A^{\frac{1}{2}}, J, L_2(\Omega)) = W_{\pm}.$$

In particular, this verifies (A.45) and (A.46).

A proof of (A.47) in the context of the Birman theory may be based on the following

LEMMA A.5. *Under the hypotheses of Theorem* A.3

$$\text{(A.52)}\qquad AW_{\pm}^{*}\Pi_0(I) = W_{\pm}^{*}A_0\Pi_0(I)$$

for every interval $I \subset \mathbb{R}$.

PROOF. Note that by (A.42), (A.43)

$$\text{(A.53)}\qquad W_{\pm}^{*} = \underset{t\to\pm\infty}{s-\lim}\, e^{itA}J^{*}e^{-itA_0}.$$

Moreover,

$$\text{(A.54)}\qquad \begin{cases} Ae^{itA}J^{*}e^{-itA_0}\Pi_0(I)u - e^{itA}J^{*}e^{-itA_0}A_0\Pi_0(I)u \\ = e^{itA}(AJ^{*}-J^{*}A_0)\Pi_0(I)e^{-itA_0}u \end{cases}$$

for all $u \in L_2(\mathbb{R}^n)$. Now $e^{-itA_0}u$ tends weakly to zero in $L_2(\mathbb{R}^n)$ when $t \to \pm\infty$, by the argument used in the proof of Theorem 5.5. But $(AJ^{*}-J^{*}A_0)\Pi_0(I) \in B_1(L_2(\mathbb{R}^n), L_2(\Omega)) \subset B_0(L_2(\mathbb{R}^n), L_2(\Omega))$. Hence, it follows as in the proof of Theorem 5.5 that the right-hand side of (A.54) tends to zero in $L_2(\Omega)$ for every $u \in L_2(\mathbb{R}^n)$. Thus passage to the limit $t \to \pm\infty$ in (A.54) gives (A.52).

PROOF OF (A.47). Note that (A.52) implies

$$(A.55)\qquad \begin{cases} A^2W_\pm^*\Pi_0(I) = A(W_\pm^*A_0\Pi_0(I)) = AW_\pm^*\Pi_0(I)A_0 \\ \qquad\qquad\quad = W_\pm^*A_0\Pi_0(I)A_0 = W_\pm^*A_0^2\Pi(I)\,. \end{cases}$$

It follows that

$$(A.56)\qquad A^m W_\pm^*\Pi_0(I) = W_\pm^* A_0^m \Pi_0(I) \quad \text{for} \quad m = 0,1,2,\ldots$$

by induction on m. Hence

$$(A.57)\qquad P(A)W_\pm^*\Pi_0(I) = W_\pm^*P(A_0)\Pi_0(I)$$

for every polynomial $P(\lambda)$. Choose a sequence $\{P_m(\lambda)\}$ of polynomials which converges monotonically to $H(\mu-\lambda)$ where μ is fixed and $H(\tau)$ is the Heaviside function. Then $P_m(A)$ converges monotonically to $H(\mu-A) = \Pi(\mu)$ [33] and similarly $P_m(A_0)$ converges monotonically to $\Pi_0(\mu)$. Thus passage to the limit in (A.57) gives

$$(A.58)\qquad \Pi(\mu)W_\pm^*\Pi_0(I) = W_\pm^*\Pi_0(\mu)\Pi_0(I) \quad \text{for all} \quad \mu \in \mathbb{R}.$$

Finally, making $I \to \mathbb{R}$ in (A.58) gives

$$(A.59)\qquad \Pi(\mu)W_\pm^* = W_\pm^*\Pi_0(\mu) \quad \text{for all} \quad \mu \in \mathbb{R}$$

which is equivalent to (A.47).

REFERENCES

1. AGMON, S., *Lectures on Elliptic Boundary Value Problems,* New York: Van Nostrand, 1965.

2. BELOPOLSKII, A. L., and M. S. BIRMAN, *The existence of wave operators in the theory of scattering with a pair of spaces,* Izv. Akad. Nauk SSSR, 32, 1162-1175 (1968) = Math. USSR Izv., *2*,1117-1130 (1968).

3. BIRMAN, M. S., *Scattering problems for differential operators with constant coefficients,* Funkcional. Anal. i. Prilozen., *3,* 1-16 (1969).

4. BIRMAN, M. S., *A criterion for the existence of the complete wave operators in the theory of scattering with two spaces,* in Topics in Mathematical Physics V. 4, M. S. Birman Ed., New York: Consultants Bureau, 1971.

5. DUNFORD, N., and J. T. SCHWARTZ, *Linear Operators I,* New York: Interscience Publishers, Inc., 1957.

6. DUNFORD, N., and J. T. SCHWARTZ, *Linear Operators II,* New York: Interscience Publishers Inc., 1963.

7. EIDUS, D. M., *The principle of limiting absorption,* Mat. Sb. 57, 13-44 (1962) = A. M. S. Transl. (2) *47*, 157-191 (1965).

8. EIDUS, D. M., *The principle of limiting amplitude,* Uspekhi Mat. Nauk, 24 (1969) = Russian Mathematical Surveys, 24, 97-167 (1969).

9. FRIEDLANDER, F. G., *Sound Pulses,* Cambridge: Cambridge University Press, 1958.

10. HALMOS, P., *Introduction to Hilbert space and the*

Theory of Spectral Multiplicity, New York: Chelsea, 1951.

11. HEINS, A. E., and S. SILVER, *The edge conditions and field representation theorems in the theory of electromagnetic diffraction*, Proc. Cam. Phil. Soc. *51*, 149-161 (1955).

12. HILLE, E., and R. S. Phillips, *Functional Analysis and Semi-Groups*, Amer. Math. Soc. Colloquium Publ., *31*, 1957.

13. HÖRMANDER, L., *Linear Partial Differential Operators*, Berlin-Göttingen-Heidelberg: Springer, 1963.

14. IKEBE, T., *Eigenfunction expansions associated with the Schrödinger operator and their application to scattering theory*, Arch. Rational Mech. Anal., *5*, 1-34 (1960).

15. IKEBE, T., *Orthogonality of the eigenfunctions for the exterior problem connected with* $-\Delta$, Arch. Rational Mech. Anal. *19*, 71-73 (1965).

16. IKEBE, T., *Remarks on the orthogonality of the eigenfunctions for the Schrodinger operator in* $\mathbb{R}^n$, J. Fac. Sci., Univ. Tokyo *17*, 355-361 (1970).

17. JÄGER, W., *Zur Theorie der Schwingungsgleichung mit variabeln Koeffizienten in Aussengebieten*, Math. Z. *102*, 62-88 (1967).

18. KATO, T. *Perturbation Theory for Linear Operators*, New York, Springer, 1966.

19. LAX, P. D., and R. S. PHILLIPS, *Scattering theory*, Bull. A. M. S., *70*, 130-142 (1964).

20. LAX, P. D., and R. S. PHILLIPS, *Scattering Theory*, New York: Academic Press, 1967.

21. LAX, P. D., and R. S. PHILLIPS, *Scattering theory for the acoustic equation in an even number of space dimensions*, Indiana Univ. Math. J., *22*, 101-134 (1972/73).

22. LIONS, J. L. and E. MAGENES, *Non-Homogeneous Boundary Value Problems and Applications* I, New York, Springer, 1972.

23. LITTMAN, W., *Fourier transforms of surface-carried measures and differentiability of surface averages*, Bull. Amer. Math. Soc. *69*, 766-770 (1963).

24. LYFORD, W. C., *Scattering theory for the Laplacian in perturbed cylindrical domains*, thesis, Calif. Inst. of Tech., June 1973.

25. LYFORD, W. C., *Scattering theory for the Laplacian in domains with cylinders*, ONR Technical Summary Rept. # 23, Univ. of Utah (March 1974).

26. MAJDA, A., *Outgoing solutions for perturbations of* $-\Delta$ *with applications to spectral and scattering theory*, preprint, Stanford Univ. (1972).

27. MAGNUS, W., F. OBERHETTINGER, and R. P. SONI, *Formulas and Theorems for the Special Functions of Mathematical Physics*, New York: Springer, 1966.

28. MATSUMURA, M., *Comportement des solutions de quelques problèmes mixtes pour certains systèmes hyperboliques symétriques à coefficients constants*, Publ. RIMS, Kyoto Univ. *4*, 309-359 (1968).

29. MATSUMURA, M., *Uniform estimates of elementary solutions of first order systems of partial differential equations*, Publ. RIMS, Kyoto Univ. *6*, 293-305 (1970).

30. MAURIN, K., *Methods of Hilbert Spaces*, New York, Hafner Publishing Co., 1967.

31. RELLICH, F., *Ein Satz über mittlere Konvergenz*, Gött. Nachr. (math. phys.), 30-35 (1930).

32. RELLICH, F., *Über das asymptotische Verhalten der Lösungen von* $\Delta u + ku = 0$ *in unendlichen Gebieten*, Jber. Deutschen Math. Verein. *53*, 57-64 (1943).

33. RIESZ, F., and B. SZ.-NAGY, *Functional Analysis*.

New York: Ungar Publishing Co., 1955.

34. SCHULENBERGER, J. R., and C. H. WILCOX, *Completeness of the wave operators for perturbations of uniformly propagative systems*, J. Functional Anal., *7*, 447-474 (1971).

35. SHENK, N. A., II, *Eigenfunction expansions and scattering theory for the wave equation in an exterior region*, Arch. Rational Mech. Anal., *21*, 120-150 (1966).

36. SHENK, N. A., II, and D. W. THOE, *Eigenfunction expansions and scattering theory for perturbations of* $-\Delta$, Rocky Mountain J. of Math. *1*, 89-125 (1971).

37. STINESPRING. W. F., *A sufficient condition for an integral operator to have a trace*, J. Reine Angew. Math. *200*, 200-207 (1958).

38. STONE, M. H., *Linear Transformations in Hilbert Space and Their Applications to Analysis*, Amer. Math. Soc. Colloquium Publ., *15*, 1932.

39. THOE, D. W., *Eigenfunction expansions associated with Shroedinger operators in* R^n, $n \geq 4$, Arch. Rational Mech. Anal., *26*, 335-356 (1967).

40. TITCHMARSH, E. C., *Introduction to the Theory of Fourier Integrals*, Oxford University Press, 1937.

41. VISHIK, M., and O. A. LADYZHENSKAYA, *Boundary value problems for partial differential equations and certain classes of operator equations*, Uspekhi Mat. Nauk *11*, 41-97 (1956) = A. M. S. Transl. (2) *10*, 223-281 (1958).

42. WERNER, P., *Zur mathematischen Theorie akustischer Wellenfelder*, Arch. Rational Mech. Anal., *6*, 231-260 (1960).

43. WILCOX, C. H., *Initial-boundary value problems for linear hyperbolic partial differential equations of the second order*, Arch. Rational Mech.

Anal., *10*, 361-400 (1962).

44. WILCOX, C. H., *Scattering states and wave operators in the abstract theory of scattering*, J. Functional Anal., *12*, 257-274 (1973).

45. WILCOX, C. H., *Asymptotic wave functions and energy distributions for the d'Alembert wave equation*, ONR Technical Summary Rept. # 24, Univ. of Utah (March 1974).

Vol. 277: Séminaire Banach. Edité par C. Houzel. VII, 229 pages. 1972. DM 22,-

Vol. 278: H. Jacquet, Automorphic Forms on GL(2) Part II. XIII, 142 pages. 1972. DM 18,-

Vol. 279: R. Bott, S. Gitler and I. M. James, Lectures on Algebraic and Differential Topology. V, 174 pages. 1972. DM 20,-

Vol. 280: Conference on the Theory of Ordinary and Partial Differential Equations. Edited by W. N. Everitt and B. D. Sleeman. XV, 367 pages. 1972. DM 29,-

Vol. 281: Coherence in Categories. Edited by S. Mac Lane. VII, 235 pages. 1972. DM 22,-

Vol. 282: W. Klingenberg und P. Flaschel, Riemannsche Hilbertmannigfaltigkeiten. Periodische Geodätische. VII, 211 Seiten. 1972. DM 22,-

Vol. 283: L. Illusie, Complexe Cotangent et Déformations II. VII, 304 pages. 1972. DM 27,-

Vol. 284: P. A. Meyer, Martingales and Stochastic Integrals I. VI, 89 pages. 1972. DM 18,-

Vol. 285: P. de la Harpe, Classical Banach-Lie Algebras and Banach-Lie Groups of Operators in Hilbert Space. III, 160 pages. 1972. DM 18,-

Vol. 286: S. Murakami, On Automorphisms of Siegel Domains. V, 95 pages. 1972. DM 18,-

Vol. 287: Hyperfunctions and Pseudo-Differential Equations. Edited by H. Komatsu. VII, 529 pages. 1973. DM 40,-

Vol. 288: Groupes de Monodromie en Géométrie Algébrique. (SGA 7 I). Dirigé par A. Grothendieck. IX, 523 pages. 1972. DM 55,-

Vol. 289: B. Fuglede, Finely Harmonic Functions. III, 188. 1972. DM 20,-

Vol. 290: D. B. Zagier, Equivariant Pontrjagin Classes and Applications to Orbit Spaces. IX, 130 pages. 1972. DM 18,-

Vol. 291: P. Orlik, Seifert Manifolds. VIII, 155 pages. 1972. DM 18,-

Vol. 292: W. D. Wallis, A. P. Street and J. S. Wallis, Combinatorics: Room Squares, Sum-Free Sets, Hadamard Matrices. V, 508 pages. 1972. DM 55,-

Vol. 293: R. A. DeVore, The Approximation of Continuous Functions by Positive Linear Operators. VIII, 289 pages. 1972. DM 27,-

Vol. 294: Stability of Stochastic Dynamical Systems. Edited by R. F. Curtain. IX, 332 pages. 1972. DM 29,-

Vol. 295: C. Dellacherie, Ensembles Analytiques Capacités. Mesures de Hausdorff. XII, 123 pages. 1972. DM 18,-

Vol. 296: Probability and Information Theory II. Edited by M. Behara, K. Krickeberg and J. Wolfowitz. V, 223 pages. 1973. DM 22,-

Vol. 297: J. Garnett, Analytic Capacity and Measure. IV, 138 pages. 1972. DM 18,-

Vol. 298: Proceedings of the Second Conference on Compact Transformation Groups. Part 1. XIII, 453 pages. 1972. DM 35,-

Vol. 299: Proceedings of the Second Conference on Compact Transformation Groups. Part 2. XIV, 327 pages. 1972. DM 29,-

Vol. 300: P. Eymard, Moyennes Invariantes et Représentations Unitaires. II, 113 pages. 1972. DM 18,-

Vol. 301: F. Pittnauer, Vorlesungen über asymptotische Reihen. VI, 186 Seiten. 1972. DM 18,-

Vol. 302: M. Demazure, Lectures on p-Divisible Groups. V, 98 pages. 1972. DM 18,-

Vol. 303: Graph Theory and Applications. Edited by Y. Alavi, D. R. Lick and A. T. White. IX, 329 pages. 1972. DM 26,-

Vol. 304: A. K. Bousfield and D. M. Kan, Homotopy Limits, Completions and Localizations. V, 348 pages. 1972. DM 29,-

Vol. 305: Théorie des Topos et Cohomologie Etale des Schémas. Tome 3. (SGA 4). Dirigé par M. Artin, A. Grothendieck et J. L. Verdier. VI, 640 pages. 1973. DM 55,-

Vol. 306: H. Luckhardt, Extensional Gödel Functional Interpretation. VI, 161 pages. 1973. DM 20,-

Vol. 307: J. L. Bretagnolle, S. D. Chatterji et P.-A. Meyer, Ecole d'été de Probabilités: Processus Stochastiques. VI, 198 pages. 1973. DM 22,-

Vol. 308: D. Knutson, λ-Rings and the Representation Theory of the Symmetric Group. IV, 203 pages. 1973. DM 22,-

Vol. 309: D. H. Sattinger, Topics in Stability and Bifurcation Theory. VI, 190 pages. 1973. DM 20,-

Vol. 310: B. Iversen, Generic Local Structure of the Morphisms in Commutative Algebra. IV, 108 pages. 1973. DM 18,-

Vol. 311: Conference on Commutative Algebra. Edited by J. W. Brewer and E. A. Rutter. VII, 251 pages. 1973. DM 24,-

Vol. 312: Symposium on Ordinary Differential Equations. Edited by W. A. Harris, Jr. and Y. Sibuya. VIII, 204 pages. 1973. DM 22,-

Vol. 313: K. Jörgens and J. Weidmann, Spectral Properties of Hamiltonian Operators. III, 140 pages. 1973. DM 18,-

Vol. 314: M. Deuring, Lectures on the Theory of Algebraic Functions of One Variable. VI, 151 pages. 1973. DM 18,-

Vol. 315: K. Bichteler, Integration Theory (with Special Attention to Vector Measures). VI, 357 pages. 1973. DM 29,-

Vol. 316: Symposium on Non-Well-Posed Problems and Logarithmic Convexity. Edited by R. J. Knops. V, 176 pages. 1973. DM 20,-

Vol. 317: Séminaire Bourbaki - vol. 1971/72. Exposés 400-417. IV, 361 pages. 1973. DM 29,-

Vol. 318: Recent Advances in Topological Dynamics. Edited by A. Beck. VIII, 285 pages. 1973. DM 27,-

Vol. 319: Conference on Group Theory. Edited by R. W. Gatterdam and K. W. Weston. V, 188 pages. 1973. DM 20,-

Vol. 320: Modular Functions of One Variable I. Edited by W. Kuyk. V, 195 pages. 1973. DM 20,-

Vol. 321: Séminaire de Probabilités VII. Edité par P. A. Meyer. VI, 322 pages. 1973. DM 29,-

Vol. 322: Nonlinear Problems in the Physical Sciences and Biology. Edited by I. Stakgold, D. D. Joseph and D. H. Sattinger. VIII, 357 pages. 1973. DM 29,-

Vol. 323: J. L. Lions, Perturbations Singulières dans les Problèmes aux Limites et en Contrôle Optimal. XII, 645 pages. 1973. DM 46,-

Vol. 324: K. Kreith, Oscillation Theory. VI, 109 pages. 1973. DM 18,-

Vol. 325: C.-C. Chou, La Transformation de Fourier Complexe et L'Equation de Convolution. IX, 137 pages. 1973. DM 18,-

Vol. 326: A. Robert, Elliptic Curves. VIII, 264 pages. 1973. DM 24,-

Vol. 327: E. Matlis, One-Dimensional Cohen-Macaulay Rings. XII, 157 pages. 1973. DM 20,-

Vol. 328: J. R. Büchi and D. Siefkes, The Monadic Second Order Theory of All Countable Ordinals. VI, 217 pages. 1973. DM 22,-

Vol. 329: W. Trebels, Multipliers for (C, α)-Bounded Fourier Expansions in Banach Spaces and Approximation Theory. VII, 103 pages. 1973. DM 18,-

Vol. 330: Proceedings of the Second Japan-USSR Symposium on Probability Theory. Edited by G. Maruyama and Yu. V. Prokhorov. VI, 550 pages. 1973. DM 40,-

Vol. 331: Summer School on Topological Vector Spaces. Edited by L. Waelbroeck. VI, 226 pages. 1973. DM 22,-

Vol. 332: Séminaire Pierre Lelong (Analyse) Année 1971-1972. V, 131 pages. 1973. DM 18,-

Vol. 333: Numerische, insbesondere approximationstheoretische Behandlung von Funktionalgleichungen. Herausgegeben von R. Ansorge und W. Törnig. VI, 296 Seiten. 1973. DM 27,-

Vol. 334: F. Schweiger, The Metrical Theory of Jacobi-Perron Algorithm. V, 111 pages. 1973. DM 18,-

Vol. 335: H. Huck, R. Roitzsch, U. Simon, W. Vortisch, R. Walden, B. Wegner und W. Wendland, Beweismethoden der Differentialgeometrie im Großen. IX, 159 Seiten. 1973. DM 20,-

Vol. 336: L'Analyse Harmonique dans le Domaine Complexe. Edité par E. J. Akutowicz. VIII, 169 pages. 1973. DM 20,-

Vol. 337: Cambridge Summer School in Mathematical Logic. Edited by A. R. D. Mathias and H. Rogers. IX, 660 pages. 1973. DM 46,-

Vol. 338: J. Lindenstrauss and L. Tzafriri, Classical Banach Spaces. IX, 243 pages. 1973. DM 24,-

Vol. 339: G. Kempf, F. Knudsen, D. Mumford and B. Saint-Donat, Toroidal Embeddings I. VIII, 209 pages. 1973. DM 22,-

Vol. 340: Groupes de Monodromie en Géométrie Algébrique. (SGA 7 II). Par P. Deligne et N. Katz. X, 438 pages. 1973. DM 44,-

Vol. 341: Algebraic K-Theory I, Higher K-Theories. Edited by H. Bass. XV, 335 pages. 1973. DM 29,-

Vol. 342: Algebraic K-Theory II, "Classical" Algebraic K-Theory, and Connections with Arithmetic. Edited by H. Bass. XV, 527 pages. 1973. DM 40,-

Vol. 343: Algebraic K-Theory III, Hermitian K-Theory and Geometric Applications. Edited by H. Bass. XV, 572 pages. 1973. DM 40,-

Vol. 344: A. S. Troelstra (Editor), Metamathematical Investigation of Intuitionistic Arithmetic and Analysis. XVII, 485 pages. 1973. DM 38,-

Vol. 345: Proceedings of a Conference on Operator Theory. Edited by P. A. Fillmore. VI, 228 pages. 1973. DM 22,-

Vol. 346: Fučík et al., Spectral Analysis of Nonlinear Operators. II, 287 pages. 1973. DM 26,-

Vol. 347: J. M. Boardman and R. M. Vogt, Homotopy Invariant Algebraic Structures on Topological Spaces. X, 257 pages. 1973. DM 24,-

Vol. 348: A. M. Mathai and R. K. Saxena, Generalized Hypergeometric Functions with Applications in Statistics and Physical Sciences. VII, 314 pages. 1973. DM 26,-

Vol. 349: Modular Functions of One Variable II. Edited by W. Kuyk and P. Deligne. V, 598 pages. 1973. DM 38,-

Vol. 350: Modular Functions of One Variable III. Edited by W. Kuyk and J. P. Serre. V, 350 pages. 1973. DM 26,-

Vol. 351: H. Tachikawa, Quasi-Frobenius Rings and Generalizations. XI, 172 pages. 1973. DM 20,-

Vol. 352: J. D. Fay, Theta Functions on Riemann Surfaces. V, 137 pages. 1973. DM 18,-

Vol. 353: Proceedings of the Conference. on Orders, Group Rings and Related Topics. Organized by J. S. Hsia, M. L. Madan and T. G. Ralley. X, 224 pages. 1973. DM 22,-

Vol. 354: K. J. Devlin, Aspects of Constructibility. XII, 240 pages. 1973. DM 24,-

Vol. 355: M. Sion, A Theory of Semigroup Valued Measures. V, 140 pages. 1973. DM 18,-

Vol. 356: W. L. J. van der Kallen, Infinitesimally Central-Extensions of Chevalley Groups. VII, 147 pages. 1973. DM 18,-

Vol. 357: W. Borho, P. Gabriel und R. Rentschler, Primideale in Einhüllenden auflösbarer Lie-Algebren. V, 182 Seiten. 1973. DM 20,-

Vol. 358: F. L. Williams, Tensor Products of Principal Series Representations. VI, 132 pages. 1973. DM 18,-

Vol. 359: U. Stammbach, Homology in Group Theory. VIII, 183 pages. 1973. DM 20,-

Vol. 360: W. J. Padgett and R. L. Taylor, Laws of Large Numbers for Normed Linear Spaces and Certain Fréchet Spaces. VI, 111 pages. 1973. DM 18,-

Vol. 361: J. W. Schutz, Foundations of Special Relativity: Kinematic Axioms for Minkowski Space Time. XX, 314 pages. 1973. DM 26,-

Vol. 362: Proceedings of the Conference on Numerical Solution of Ordinary Differential Equations. Edited by D. Bettis. VIII, 490 pages. 1974. DM 34,-

Vol. 363: Conference on the Numerical Solution of Differential Equations. Edited by G. A. Watson. IX, 221 pages. 1974. DM 20,-

Vol. 364: Proceedings on Infinite Dimensional Holomorphy. Edited by T. L. Hayden and T. J. Suffridge. VII, 212 pages. 1974. DM 20,-

Vol. 365: R. P. Gilbert, Constructive Methods for Elliptic Equations. VII, 397 pages. 1974. DM 26,-

Vol. 366: R. Steinberg, Conjugacy Classes in Algebraic Groups (Notes by V. V. Deodhar). VI, 159 pages. 1974. DM 18,-

Vol. 367: K. Langmann und W. Lütkebohmert, Cousinverteilungen und Fortsetzungssätze. VI, 151 Seiten. 1974. DM 16,-

Vol. 368: R. J. Milgram, Unstable Homotopy from the Stable Point of View. V, 109 pages. 1974. DM 16,-

Vol. 369: Victoria Symposium on Nonstandard Analysis. Edited by A. Hurd and P. Loeb. XVIII, 339 pages. 1974. DM 26,-

Vol. 370: B. Mazur and W. Messing, Universal Extensions and One Dimensional Crystalline Cohomology. VII, 134 pages. 1974. DM 16,-

Vol. 371: V. Poenaru, Analyse Différentielle. V, 228 pages. 1974. DM 20,-

Vol. 372: Proceedings of the Second International Conference on the Theory of Groups 1973. Edited by M. F. Newman. VII, 740 pages. 1974. DM 48,-

Vol. 373: A. E. R. Woodcock and T. Poston, A Geometrical Study of the Elementary Catastrophes. V, 257 pages. 1974. DM 22,-

Vol. 374: S. Yamamuro, Differential Calculus in Topological Linear Spaces. IV, 179 pages. 1974. DM 18,-

Vol. 375: Topology Conference 1973. Edited by R. F. Dickman Jr. and P. Fletcher. X, 283 pages. 1974. DM 24,-

Vol. 376: D. B. Osteyee and I. J. Good, Information, Weight of Evidence, the Singularity between Probability Measures and Signal Detection. XI, 156 pages. 1974. DM 16,-

Vol. 377: A. M. Fink, Almost Periodic Differential Equations. VIII, 336 pages. 1974. DM 26,-

Vol. 378: TOPO 72 - General Topology and its Applications. Proceedings 1972. Edited by R. Alò, R. W. Heath and J. Nagata. XIV, 651 pages. 1974. DM 50,-

Vol. 379: A. Badrikian et S. Chevet, Mesures Cylindriques, Espaces de Wiener et Fonctions Aléatoires Gaussiennes. X, 383 pages. 1974. DM 32,-

Vol. 380: M. Petrich, Rings and Semigroups. VIII, 182 pages. 1974. DM 18,-

Vol. 381: Séminaire de Probabilités VIII. Edité par P. A. Meyer. IX, 354 pages. 1974. DM 32,-

Vol. 382: J. H. van Lint, Combinatorial Theory Seminar Eindhoven University of Technology. VI, 131 pages. 1974. DM 18,-

Vol. 383: Séminaire Bourbaki - vol. 1972/73. Exposés 418-435 IV, 334 pages. 1974. DM 30,-

Vol. 384: Functional Analysis and Applications, Proceedings 1972. Edited by L. Nachbin. V, 270 pages. 1974. DM 22,-

Vol. 385: J. Douglas Jr. and T. Dupont, Collocation Methods for Parabolic Equations in a Single Space Variable (Based on C^1-Piecewise-Polynomial Spaces). V, 147 pages. 1974. DM 16,-

Vol. 386: J. Tits, Buildings of Spherical Type and Finite BN-Pairs. IX, 299 pages. 1974. DM 24,-

Vol. 387: C. P. Bruter, Eléments de la Théorie des Matroïdes. V, 138 pages. 1974. DM 18,-

Vol. 388: R. L. Lipsman, Group Representations. X, 166 pages. 1974. DM 20,-

Vol. 389: M.-A. Knus et M. Ojanguren, Théorie de la Descente et Algèbres d' Azumaya. IV, 163 pages. 1974. DM 20,-

Vol. 390: P. A. Meyer, P. Priouret et F. Spitzer, Ecole d'Eté de Probabilités de Saint-Flour III - 1973. Edité par A. Badrikian et P.-L. Hennequin. VIII, 189 pages. 1974. DM 20,-

Vol. 391: J. Gray, Formal Category Theory: Adjointness for 2-Categories. XII, 282 pages. 1974. DM 24,-

Vol. 392: Géométrie Différentielle, Colloque, Santiago de Compostela, Espagne 1972. Edité par E. Vidal. VI, 225 pages. 1974. DM 20,-

Vol. 393: G. Wassermann, Stability of Unfoldings. IX, 164 pages. 1974. DM 20,-

Vol. 394: W. M. Patterson 3rd, Iterative Methods for the Solution of a Linear Operator Equation in Hilbert Space - A Survey. III, 183 pages. 1974. DM 20,-

Vol. 395: Numerische Behandlung nichtlinearer Integrodifferential- und Differentialgleichungen. Tagung 1973. Herausgegeben von R. Ansorge und W. Törnig. VII, 313 Seiten. 1974. DM 28,-

Vol. 396: K. H. Hofmann, M. Mislove and A. Stralka, The Pontryagin Duality of Compact O-Dimensional Semilattices and its Applications. XVI, 122 pages. 1974. DM 18,-

Vol. 397: T. Yamada, The Schur Subgroup of the Brauer Group. V, 159 pages. 1974. DM 18,-

Vol. 398: Théories de l'Information, Actes des Rencontres de Marseille-Luminy, 1973. Edité par J. Kampé de Fériet et C. Picard. XII, 201 pages. 1974. DM 23,-